T0339649

Phytochemical Investigations of Genus *Terminalia*

Phytochemical Investigations of Medicinal Plants

Series Editor:
Brijesh Kumar

Phytochemistry of Plants of Genus *Phyllanthus*
Brijesh Kumar, Sunil Kumar and K. P. Madhusudanan

Phytochemistry of Plants of Genus *Ocimum*
Brijesh Kumar, Vikas Bajpai, Surabhi Tiwari and Renu Pandey

Phytochemistry of Plants of Genus *Piper*
Brijesh Kumar, Surabhi Tiwari, Vikas Bajpai and Bikarma Singh

Phytochemistry of *Tinospora cordifolia*
Brijesh Kumar, Vikas Bajpai and Nikhil Kumar

Phytochemistry of Plants of Genus *Rauvolfi*a
Brijesh Kumar, Sunil Kumar, Vikas Bajpai and K. P. Madhusudanan

Phytochemistry of *Piper betle* Landacres
Vikas Bajpai, Nikhil Kumar and Brijesh Kumar

Phytochemical Investigations of Genus *Terminalia*
Brijesh Kumar, Awantika Singh, and K. P. Madhusudanan

Phytochemistry of plants of Genus *Cassia*
Brijesh Kumar, Vikas Bajpai, Vikaskumar Gond, Subhashis Pal, and Naibedya Chattopadhyay

For more information about this series, please visit: https://www.crcpress.com/
Phytochemical-Investigations-of-Medicinal-Plants/book-series/PHYTO

Phytochemical Investigations of Genus *Terminalia*

Brijesh Kumar, Awantika Singh, and
K. P. Madhusudanan

CRC Press
Taylor & Francis Group
Boca Raton London New York

CRC Press is an imprint of the
Taylor & Francis Group, an **informa** business

First edition published 2022
by CRC Press
6000 Broken Sound Parkway NW, Suite 300, Boca Raton, FL 33487-2742

and by CRC Press
2 Park Square, Milton Park, Abingdon, Oxon, OX14 4RN

©2022 selection and editorial matter, Brijesh Kumar, Awantika Singh and K. P. Madhusudanan; individual chapters the contributors.

The right of Brijesh Kumar, Awantika Singh and K. P. Madhusudanan to be identified as the authors of the editorial material, and of the authors for their individual chapters, has been asserted in accordance with sections 77 and 78 of the Copyright, Designs and Patent Act 1988.

CRC Press is an imprint of Taylor & Francis Group, LLC

ISBN: 978-1-032-01948-2 (hbk)
ISBN: 978-1-032-01949-9 (pbk)
ISBN: 978-1-003-18112-5 (ebk)

Typeset in Times
by codeMantra

Contents

List of figures

List of tables

Preface

Plants of the genus *Terminalia* are among the most widely used for traditional medicinal purposes. *Terminalia* species are distributed in the tropical and subtropical regions of Asia, Australia, Africa and America. The most common species of *Terminalia* found in India are *Terminalia arjuna*, *Terminalia bellirica*, *Terminalia catappa*, *Terminalia chebula*, *Terminalia elliptica* and *Terminalia paniculata*. They are found in the sub-Himalayan region, the plains, hilly regions and the coastal regions. All are deciduous trees growing up to a height of around 30 m. Plants belonging to the genus *Terminalia* have a long history of use in traditional medicinal systems, and they are extensively used in several continents for the treatment of cardiovascular diseases, wound healing, abdominal disorders, bacterial infections, cold, sore throat, conjunctivitis, diarrhea, dysentery, fever, gastric ulcers, headaches, heart diseases, hookworm infestation, hypertension, jaundice, leprosy, nosebleed, edema, pneumonia and skin diseases. The phytochemistry, phytochemical analysis and pharmacology are reviewed.

A simple, reliable and reproducible method for screening the phytochemicals of *Terminalia* species (*T. arjuna*, *T. bellirica*, *T. catappa*, *T. chebula*, *T. elliptica* and *T. paniculata*) was developed using HPLC-ESI-QTOF-MS/MS. Different classes of compounds such as acids, triterpenoids, flavonoids, tannins/polyphenols such as gallic and ellagic acid derivatives and proanthocyanidins have been identified and characterized from the plant extracts by their RT, exact mass measurement, isotopic peak pattern, molecular formula and MS/MS fragmentation patterns. A total of 179 compounds including 25 acids/triterpenoids, 35 flavonoids, 27 gallic acid derivatives, 49 ellagic acid derivatives and 43 proanthocyanidins were tentatively identified from various parts of the six *Terminalia* species.

Furthermore, a robust and sensitive UPLC-ESI-QqQ$_{LIT}$-MS/MS method under MRM mode was developed and validated for the simultaneous quantification of 37 bioactive compounds in different parts of the six *Terminalia* species and 17 bioactive compounds in *Terminalia chebula* fruits collected from different locations and its polyherbal formulations. The results indicated that the total contents of all the 37 compounds were abundant in *T. chebula* fruit. Discrimination of different plant parts of the six *Terminalia* species was effectively done by PCA. The quantitative results also indicated the importance of plant parts, and comprehensive estimation of selected bioactive compounds could contribute to quality control of *Terminalia* species and its related preparations. A wide variation of the contents of the 17 bioactive compounds in *T. chebula* fruits from different locations indicated the effect of geographic location on the contents. The polyherbal formulations investigated also showed wide variations and significant differences in the contents pointing out inadequate standardization. Comprehensive estimation of selected bioactive compounds could contribute to quality control of *T. chebula* species and its herbal formulations.

Acknowledgments

The completion of this book is due to the Almighty who blessed us with all the resources required to accomplish this journey. We are glad to have this chance to express our gratitude to people who have been supportive to us every time. We express our deep sense of gratitude to the Director, CSIR-Central Drug Research Institute (CDRI), Lucknow for his support and Sophisticated Analytical Instrument Facility (SAIF) Division, CSIR-CDRI, India where all the data were generated.

Authors

Dr. Brijesh Kumar

Dr. Brijesh Kumar is a Professor (AcSIR) and Chief Scientist of the Sophisticated Analytical Instrument Facility Division, CSIR-Central Drug Research Institute Lucknow, India. Currently, he is facility in-charge of the Sophisticated Analytical Instrument Facility (SAIF). He has completed his PhD from CSIR-CDRI, Lucknow (Dr. R.M.L Avadh University, Faizabad, UP, India). He has to his credit 11 book chapters, 7 books and 150 papers in international journals of repute. His current area of research is applications of mass spectrometry (DART MS/Q-TOF LC-MS/4000 Q Trap LC-MS/Orbitrap MS^n) for qualitative and quantitative analyses of phytoconstituents of Indian medicinal plants and their herbal formulations for quality control and authentication/standardization. He is also involved in the identification of marker compounds using statistical software to check adulteration/substitution.

Dr. Awantika Singh

Awantika Singh completed her PhD from the Academy of Scientific and Innovative Research (AcSIR), New Delhi, India and carried out her research work under the supervision of Dr. Brijesh Kumar at CSIR-Central Drug Research Institute Lucknow. Her research includes application of hyphenated mass spectrometric techniques for qualitative and quantitative analyses of phytochemicals from selected Indian medicinal plants.

Dr. K. P. Madhusudanan

Dr. K. P. Madhusudanan is a mass spectrometry scientist born in 1947 in Kerala, India. He obtained his doctoral degree in 1975 specializing in Organic Mass Spectrometry from the National Chemical Laboratory, Pune, India. He worked as a scientist and Head of Sophisticated Analytical Instrument Facility in Central Drug Research Institute, Lucknow until 2007. His research experience since 1970 includes various aspects of organic mass spectrometry such as fragmentation mechanism, gas-phase unusual reactions, positive and negative ion mass spectrometry of natural products using various ionization techniques including DART, effects of metal cationization, LC/MS and MS/MS applications and quantitative analysis of drugs and metabolites. He has authored more than 150 research publications. He was a member of the editorial board of the *Journal of Mass Spectrometry* during 1995–2007. He is a fellow of the National Academy of Sciences, Allahabad, India. At present, he lives in Kochi.

List of abbreviations and units

°C	Degree Celsius
μg	Microgram
μL	Microliter
BPC	Base peak chromatogram
CE	Collision energy
CID	Collision-induced dissociation
CXP	Cell exit potential
Da	Dalton
DAD	Diode array detection
DP	Declustering potential
EP	Entrance potential
ESI	Electrospray ionization
eV	Electron volt
FIA	Flow injection analysis
FWHM	Full width at half maximum
g	Gram
GC-MS	Gas chromatography-mass spectrometry
GS1	Nebulizer gas
GS2	Heater gas
h	Hour
HPLC	High-performance liquid chromatography
ICH	International Conference on Harmonization
IT	Ion trap
kPa	Kilopascal
L	Liter
LC	Liquid chromatography
LOD	Limit of detection
LOQ	Limit of quantification
m	Meter
m/z	Mass-to-charge ratio
mg	Milligram

min	Minute
mL	Milliliter
MRM	Multiple reaction monitoring
MS	Mass spectrometry
ms	Millisecond
MS/MS	Tandem mass spectrometry
ng	Nanogram
NMR	Nuclear magnetic resonance
PCA	Principal component analysis
PDA	Photodiode array detection
psi	Pressure per square inch
QqQ$_{LIT}$	Hybrid linear ion trap triple quadrupole
QTOF	Quadrupole time of flight
R^2	Correlation coefficient
RDA	Retro-Diels–Alder
RSD	Relative standard deviation
S/N	Signal-to-noise ratio
SD	Standard deviation
RT	Retention time
UPLC	Ultra performance liquid chromatography
UV	Ultraviolet
XIC/EIC	Extracted ion chromatogram

Terminalia Ethno- and Phytopharmacological Review

1

1.1 INTRODUCTION

Plants and plant extracts form the basis of affordable and easily accessible sources of treatment for over three-quarters of the world's population. According to ancient literature, the therapeutic use of plants is >5,000 years old (Pan et al. 2014). There are about 374,000 plant species in the world (Christenhusz and Byng 2016). Herbal medicine (HM) or phytomedicine refers to herbs, herbal materials, herbal preparations, and finished herbal products with therapeutic and healing properties. It is the oldest and most widely used system of medicine in the world. One of the oldest forms of medicine that originated from India is the 5,000-year-old Ayurveda. It is still commonly practiced in India with 85% of Indians using crude plant preparations for a wide variety of diseases (Kamboj 2000; Prakash et al. 2017).

Almost 90% of Ayurvedic medicines are plant based. The Indian subcontinent accounts for nearly 10,000 plants used for medicinal purposes. Out of this, only 1,200–1,500 plants have been incorporated into the Ayurvedic Pharmacopia (Kumar et al. 2017; Joshi et al. 2017).

Because of their varied medicinal properties, the plants of the genus *Terminalia* are among the most widely used for traditional medicinal purposes. The genus *Terminalia*, the second largest genus of the flowering plant family Combretaceae and one of the most useful genera of therapeutic plants, has around 200 species ranging from small shrubs or trees to large deciduous forest trees distributed in the tropical and subtropical regions of the world (McGaw et al. 2001). The leaves of the plant appear at the very tips of the shoots, and thus, the genus gets its name *Terminalia* from the Latin term 'terminalis' (ending). Some species are of commercial importance for products such as gums, resins and tanning extracts as well as woods used for cabinetwork, tools and boat construction (Smith et al. 2004). Several species are traditionally used as medicines for various ailments.

Terminalia species are distributed in the tropical and subtropical regions of Asia, Australia, Africa and America. The Asian *Terminalias* covering many of the useful species with extensive documentation of their therapeutic effects occur across Malaysia, Indonesia, South Asia, West Asia and the Middle East. *Terminalia* is represented in India by 14 species out of which 6 are recorded in codified and non-codified Indian systems of medicine. The most common species of *Terminalia* found in India are *Terminalia arjuna* (Roxb. ex. DC) Wight & Arn., *Terminalia bellirica* (Gaertner) Roxb., *Terminalia catappa* L., *Terminalia chebula* Retz., *Terminalia elliptica* Willd. and *Terminalia paniculata* Roth. *T. arjuna* and *T. chebula* are well documented due to their countless uses as Ayurvedic medicines.

T. arjuna is an exotic tree in India having a wide spectrum of biological activity. *T. arjuna*, known as white marudah and regarded as a sacred medicinal plant, is an important Indian HM used as a cardioprotective agent and against various pathological infections. *T. arjuna*, indigenous to India, is a deciduous, usually evergreen, large-sized fluted tree growing up to 30 m with oval crown and drooping branches having fruits equally five-winged and distributed throughout the greater part of India, Burma and Sri Lanka. It grows in almost every part of India, but profusely in the Western Ghats, central India and sub-Himalayan region along banks of rivers, ravines and streams (Kirtikar and Basu 1989). *T. bellirica* or beleric or bastard myrobalan is a large deciduous tree growing up to 50 m having fruits not winged and distributed in semi-evergreen and moist deciduous forests and in the plains throughout India except Jammu and Kashmir, Himachal Pradesh, Sikkim and Arunachal Pradesh. It is also grown as an avenue tree. *T. catappa* (tropical almond, Indian almond or desi almond) is a fast-growing deciduous or semi-evergreen erect and spreading tree growing up to 35 m having fruits with no wings but with two ridges and distributed throughout Indian Ocean, Tropical Asia and into the Pacific Ocean (Thomson and Evans 2006). In India, it is seen on the tropical coastal environments and on roadsides as shade or ornamental trees.

T. chebula commonly known as black myrobalan is native to South Asia from India and Nepal east to southwest China. It is a medium- to large-sized deciduous tree having round crown and spreading branches growing up to a height of 40 m having fruits not winged and distributed in the sub-Himalayan regions up to an altitude of 1,500 m, northeast India and peninsular India in dry slopes up to 900 m (Bag et al. 2013). *T. elliptica* Willd. (Syn. *Terminalia tomentosa* Wight & Arn; *Terminalia alata* Heyne ex Roth; *Terminalia coriacea*

(Roxb.) Wight & Arn) or Indian Laurel is a large deciduous tree 20–35 m tall having fruits equally five-winged growing in the humid regions of India including the sub-Himalayan tracts and peninsular regions up to an altitude of 1,000 m (Khare 2007). *T. paniculata* Roth. or flowering marudah is a deciduous semi-evergreen tree growing up to 33 m at an altitude of 800–1,200 m in peninsular India and having a natural distribution in Western Ghats. It is a tropical tree having fruits unequally three-winged. The tree can be identified by its scissored and cracked bark for which reason it is sometimes known as the crocodile bark tree.

1.2 TRADITIONAL USES AND MEDICINAL PROPERTIES

Plants belonging to the genus *Terminalia* have a long history of use in traditional medicinal systems, and they are extensively used in several continents for the treatment of cardiovascular diseases, wound healing, abdominal disorders, bacterial infections, cold, sore throat, conjunctivitis, diarrhea, dysentery, fever, gastric ulcers, headaches, heart diseases, hookworm infestation, hypertension, jaundice, leprosy, nosebleed, edema, pneumonia and skin diseases (Cock 2015; Fahmy et al. 2015). Traditionally, *Terminalia* species have been used for a broad range of medicinal purposes (Table 1.1). These are well documented in Ayurveda and other Indian systems of medicine (Bharani et al. 1995; Wright et al. 2016a). *Terminalia* species are used by Ayurvedic practitioners for a variety of medicinal purposes, including abdominal and back pain, coughs and colds, conjunctivitis, diarrhea and dysentery, fever, headache, heart disorders, inflammation, leprosy, pneumonia, sexually transmitted diseases, worms, wounds, hemorrhages, ulcers, and as a general tonic. Among the Indian *Terminalia*, *T. arjuna*, *T. bellirica*, *T. chebula* and *T. catappa* are the most useful, having multiple medicinal applications (Table 1.1). They have diuretic and cardiotonic properties (Gangopadhyay and Chakrabarty 1997).

   T. arjuna is an exotic tree in India having a wide spectrum of biological activity. Its bark is mainly used for medicinal purposes, and it is antidysenteric, antipyretic, astringent, cardiotonic, demulcent, expectorant, lithotriptic, anticoagulant, hypolipidemic, antimicrobial and antiuremic and has shown to be useful in fracture, ulcers, leukorrhea, diabetes, anemia, cardiopathy and cirrhosis (Mandal et al. 2013; Warrier et al. 1993). Arjuna is a very reputed drug for all types of heart diseases and is used in the Ayurveda in the form of *asava*, *ghrita*, *kshirpaka*, powder and in various other ways. This Ayurvedic

TABLE 1.1 Traditional medicinal uses of *Terminalia* species

SPECIES	PART/HERBAL PREPARATION/ TRADITIONAL USES	MEDICINAL PROPERTIES	CONSTITUENTS
Terminalia arjuna Syn: *T. angustifolia* Common name: Arjuna, white marudah, Arjuna Myrobalan	Bark, leaf, fruit Arishta, ghritam or powder Heart diseases, asthma Bark juice is tonic, leaf juice for earache	Analgesic, antihypertensive, antiinflammatory, antioxidant, astringent, cardiac stimulant, cardioprotective, lipid lowering	Arjunetin, arjunolic acid, arjungenin, chebulinic acid, ellagic and gallic acid, flavonoids, proanthocyanidins, sitosterol, tannins, triterpenoids
Terminalia bellirica Common name: Beheda, beleric or bastard myrobalan	Bark, fruit Rasayan, fruit powder is one of the ingredients of Triphala Rejuvenative, fruit paste for wound dressing, seed oil for rheumatism, bark gum demulcent	Analgesic, alopecia, anthelmintic, antipyretic, astringent, aphrodisiac, diuretic, expectorant laxative, treatment of coughs and colds, anemia	Anolignan B belliricanin, ellagic acid, gallic acid 7-hydroxy-3,4-(methylene dioxy)-flavone, lignans sitosterol, termilignan, (-)-thannilignan
Terminalia catappa Common name Jungli badam Tropical almond Malabar almond Umbrella tree	Bark, fruit, leaf Leaf juice for leprosy and scabies. Dysentery, colic, dressing of rheumatic joints, fever, thrush, sores and abscesses. For tanning leather and dyeing textiles. Leaves for fish breeding. Water-resistant wood for canoes. Seed oil for cosmetics	Antioxidant, anticancer, for diarrhea and dysentery, liver activity against chloroquine-resistant *Plasmodium falciparum*	Flavonoids kaempferol, quercetin, phytosterols saponins and tannins

(Continued)

TABLE 1.1 (Continued)   Traditional medicinal uses of *Terminalia* species

SPECIES	PART/HERBAL PREPARATION/ TRADITIONAL USES	MEDICINAL PROPERTIES	CONSTITUENTS
Terminalia chebula Common name: Haritaki, Chebulic Myrobalan, Inknut	Bark and fruit Bark is diuretic, cardiotonic, cuts and wounds, eczema. Fruit is astringent, laxative, tonic, fruit powder one of the ingredients of Triphala. For tanning leather and dyeing cloth	Antihelmintic, antitumor, aphrodisiac cardiotonic, homeostatic, for coughs and colds, digestive disorders, eye disorders, tonic, For detoxification.	Chebulinic acid, chebulagic acid, corilagin, punicalagin, terflavin B, termilignan
Terminalia elliptica Syn; *T. tomentosa, T. alata, T. coreacea* Common name: Indian Laurel Raktarjun	Bark is astringent, antiseptic. Bark juice for cuts and wounds. Bark and fruit yield tannins for dyeing and tanning leather. Gum as an incense and cosmetic. Leaves used in sericulture. Commercial source of timber.	Astringent, antiseptic, antiinflammatory. For diarrhea and dysentery, ulcers, redness of the eye, stomachache, fever, urinary diseases	Catechol, oxalic acid, pyrogallol, flavonoids, stigmasterol, β-sitosterol, procyanidins
Terminalia paniculata Common name: Flowering marudah, Ashwakarna, Kindal	Bark, flower and leaf for cholera, cough, bronchitis, cardiac debility, wounds, skin diseases. Used in sericulture. Commercial source of timber. Substitute for teak	Flower juice and bark of *Terminalia paniculata* have been used as a remedy for cholera, diabetes, inflamed parotid glands and menstrual disorders, hemostatic	Flavonoids, ellagic and gallic acid, phenols, terminic acid, β-sitosterol, triterpenoids

remedy has been mentioned since the Vedic period in many ancient Indian medicinal texts including Charaka Samhita, Sushruta Samhita and Ashtanga Hridayam. Vagabhatta advocated the use of stem bark powder in heart ailments (Dwivedi and Chopra 2014). Being used in Ayurveda since ancient times, the dried bark of *T. arjuna* is a cardiac stimulant and cardiotonic HM for cardiovascular disorders and is used for lipid-lowering activities, alleviating angina, treatment of spermatorrhea, healing fractures, leukorrhea, diabetes and many other diseases (Dwivedi and Udupa 1989; Dwivedi 2007; Warrier et al. 1993). It is also used as an Ayurvedic medicine for urinary disorders, high blood pressure and ulcers. Its leaf juice is used to cure dysentery and earache. It has lipid-lowering and antioxidant properties similar to vitamin E. The bark is acrid, astringent and tonic with no side effects on prolonged use. A decoction of *T. arjuna* bark made with milk is taken every morning on an empty stomach to keep the heart healthy. The bark is also useful in the treatment of asthma. The decoction of the herb is used as an astringent for cleaning sores and ulcers.

T. bellirica is a powerful rejuvenating herb that nourishes the lungs, throat, voice, eyes and hairs. It is laxative and astringent purging bowels simultaneously toning the digestive tract tissues. The bark of *T. bellirica* is mildly diuretic and is useful in anemia and leucoderma, whereas its fruits are antiinflammatory, antihelmintic, expectorant, antipyretic and antiemetic and useful in asthma and bronchitis, dropsy, dyspepsia, cardiac disorders, skin diseases, leprosy and ulcer. A decoction of the bark is useful in anemia and leucoderma. Its fruit powder is one of the three ingredients in the Ayurvedic medicine "Triphala". *T. bellirica* fruits are widely used in the treatment of diarrhea, dysentery, headache, fevers, cough, tuberculosis, flatulence, leukorrhea, liver diseases, digestive disorders and gastrointestinal complaints. Application of belleric fruit paste or seed oil relieves swelling, and the fruit powder is used for wound dressing as it arrests bleeding. It also helps in lowering cholesterol and blood pressure (Deb et al. 2016). Ayurvedic practitioners use it for the treatment of hypertension, rheumatism and diabetes (Modak et al. 2007). The bark of *T. bellirica* is used as an adulterant to the bark of *T. arjuna* (Meena et al. 2010).

The fruit, flesh and seed of *T. catappa* are edible. The bark and leaves of *T. catappa* contain tannin or black dye used for tanning leather and dyeing textiles. The tannin is also used medicinally to treat rheumatism, headache, colic and diarrhea, skin eruptions and a host of others. The leaves, bark and fruit of *T. catappa* L. have been used as antidiarrheal, antipyretic, diaphoretic, cardiotonic and hemostatic, for example, to stop bleeding during tooth extraction (Lin 1992). The aqueous extract of the bark is used to treat dysentery and diarrhea (Raju and Reddy 2005). A decoction of the leaves of *T. catappa* has been used in local traditional medicine for the treatment of liver ailments, headache, colic and as a cardiotonic and diuretic (Parrotta 2001). Traditionally, tea made out of fallen leaves of *T. catappa* is used for the treatment of liver diseases, while young leaves are used for colic (Kirtikar and Basu 1991; Corner 1997). Scabies, leprosy wounds and other skin diseases are treated using a plaster of *T. catappa* leaves (Nair and Chanda 2008). *T. catappa* seed oil is used in cosmetic, pharmaceutical and food products (Monnet et al. 2012).

The dried ripe fruit of *T. chebula* is traditionally used as a homeostatic, antitussive, laxative, diuretic and cardiotonic agent, and to treat chronic ulcers and wounds. Triphala, a popular Ayurvedic formulation used as a detoxifying agent of the colon and for the treatment of stomach disorders, constipation, piles, leprosy and fever as well as various other diseases, contains fruits of *T. chebula, T. bellirica* and *E. officinalis* (Kirtikar and Basu 1935; Tarasiuk et al. 2018). It is traditionally used to prevent diseases and treat a number of symptoms, infections and inflammation, to maintain bowel health, to detoxify the body and to support the immune system (Baliga et al. 2012). It is also rejuvenative, laxative (unripe), purgative, stomachic, astringent (ripe), anthelmintic, expectorant, tonic, carminative and appetite stimulant. It is used to treat skin disorders, anemia, piles, fever, heart disease, anorexia, diarrhea, cough and excessive mucus secretion. The pulp of the fruit is used to treat piles, chronic diarrhea, dysentery, flatulence, asthma, urinary disorder, vomiting, hiccup, intestinal worms and enlarged spleen and liver (Thomas et al. 2000). Unripe fruit is used to control dysentery and diarrhea, whereas fruit powder is used to treat digestive disorders and roasted fruit is given to chew for relief from cough. Its extraordinary power of healing takes it to the top of the list in Ayurvedic Materia Medica (Bag et al. 2013). Powdered fruit made into a paste with jaggery removes impurities and wastes from the body. *T. chebula* fruit is an effective and safe purgative when taken as a powder, but a decoction of the dried fruit is used for treating diarrhea and dysentery. It improves digestion, promotes the absorption of nutrients and regulates colon function. It is useful in chronic diarrhea, chronic cough, kidney stones, dysuria and skin disorders (Asolkar et al. 1992; Bag et al. 2013). It is regarded as a rejuvenating agent and a universal panacea in Ayurveda. It is also widely used as a traditional medicine in Iran (Jokar et al. 2016).

T. elliptica has many medicinal properties like antifungal, antioxidant, antihyperglycemic, antidiarrheal and antileukorrheal properties. The bark and the fruit yield tannin, which has antiseptic and antioxidant properties. The bark is astringent and is used to treat diarrhea, burns, wounds and swellings. Tannin from the bark is used in the leather industry, while the leaves are used in sericulture. The gum exudate from the trunk is used as an incense and cosmetic and to treat stomachache. The bark is bitter and styptic, useful to treat ulcers, inflammatory conditions, fractures, hemorrhages, bronchitis cardiopathy, strangury, wounds, hemoptysis, dysentery, cough, verminosis, leukorrhea, gonorrhea and burning sensation (Kirtikar and Basu 1991; Warrier et al. 1993). *T. elliptica* bark decoction is used to treat rheumatism, fever and urinary diseases. It is a storehouse of potable water (pale yellow) issuing from the cut end of *T. elliptica* water blister. This water tastes slightly bitter and salty, but it is believed to contain amazing healing properties and is a remedy for stomachache and redness of the eye due to conjunctivitis. The timber is widely used for making furniture, the bark is a source of tannin used in the leather industry and the leaves are used in sericulture. The astringent and bitter bark of *T. paniculata* is also a source of tannin, whereas the fruit is a source of tannin and dye. Traditionally, the flower juice and bark of *T. paniculata* have been used as a remedy for cholera, diabetes, inflamed parotid glands, menstrual disorders, bronchitis, cardiac debility and diabetes (Nadkarni 1996; Srinivasan et al. 2016). It acts as a hemostatic agent. Its external application helps in wound healing and early reunion of fractures and reducing inflammation. It is antidiabetic, antibacterial, antiinflammatory and diuretic. Traditionally, it is also used to treat cough, bronchitis, cardiac debility, hepatitis, diabetes, strangury, leprosy and skin diseases (Warrier et al. 1993). *T. paniculata* is economically important for wood, medicinal uses, and raising silkworms. The wood is very hard and used as a substitute for teak.

1.3 PHYTOCHEMICAL CONSTITUENTS

Genus *Terminalia* is known to be a rich source of secondary metabolites, mainly polyphenols and triterpenoids. About 39 species have been phytochemically studied leading to the identification of 368 compounds (Zhang et al. 2019). Flavonoids and hydrolyzable tannins (gallotannins and ellagitannins) are predominant among the polyphenols, whereas oleanolic acid derivatives are the major constituents of the triterpenoid class of compounds (Cock 2015; Fahmy et al. 2015; Pfundstein et al. 2010). About 82 tannins from the genus *Terminalia* were reviewed (Chang et al. 2019). The major phytoconstituents of *T. arjuna*, *T. bellirica*, *T. catappa*, *T. chebula*, *T. elliptica* and *T. paniculata* are given in Table 1.2.

Several reviews discuss the phytochemical compositions of *T. arjuna* (Fahmy et al. 2015; Cock 2015; Jain et al. 2009; Paarakh 2010; Hafiz et al. 2014; Amalraj and Gopi 2017; Arumugam and Gopinath 2011; Gaikwad and Jadhav 2018), *T. bellirica* (Belapurkar et al. 2014; Kumar and Khurana 2018; Kumari et al. 2017), *T. catappa* (Anand et al. 2015; Fahmy et al. 2015; Dwivedi et al. 2016) and *T. chebula* (Bag et al. 2013; Walia and Arora 2013; Belapurkar et al. 2014; Riaz et al. 2017; Singh and Malhotra 2017; Upadhyay et al. 2014; Promila and Madan 2018; Chattopadhyay and Bhattacharyya 2007).

T. arjuna is a source of tannins, pseudotannins, gallic and ellagic acids, and their derivatives and non-chebulic ellagitannins. Gallic acid, ethyl gallate, 2,3,4,6-tetra-O-galloyl-β-D-Glc, and 1,2,3,4,6-penta-O-galloyl-β-D-Glc are the gallic acid derivatives reported from *T. arjuna*. Non-chebulagic ellagic tannins reported are arjunin, punicalin, punicalagin, 2,3:4,6-bis-O-HHDP-1-O-galloyl-β-D-Glc,

TABLE 1.2 Phytoconstituents of *Terminalia* species*

SL. NO.	COMPOUND	T. ARJUNA	T. BELLIRICA	T. CATAPPA	T. CHEBULA	T. ELLIPTICA/T. ALATA	T. PANICULATA
	Phenolic Acids						
1	Caffeic acid	L, S	L, S	L, SB	L, S, R, F	B	B, L, S, R, F
2	Chlorogenic acid	L, S		L, F	S, R, F, L	AP, B	L
3	m-Coumaric acid			L			
4	p-Coumaric acid			F, L	WP, Se, L		
5	3,4-Dihydroxybenzoic acid			L			
6	Ferulic acid	B, L, S, F	B, L, S, R, F	F, L, S, R	B, L, S, R, F	B, L, S, R, F	B, L, S, R, F
7	Gentisic acid			L			
8	4-Hydroxybenzoic acid			L			
9	p-Hydroxyphenylpropionic acid			L			
10	Melilotic acid				Se		
11	Protocatechuic acid	B, L, S, R, F	B, L, S, R, F	B, L, S, R, F	B, L, S, R, F	B, L, S, R, F	B, L, S, R, F
12	Quinic acid	B, L, S, R, F	B, L, S, R, F	B, L, S, R, F	B, L, S, R, F	B, L, S, R, F	B, L, S, R, F
13	Sinapic acid	B, L, S, R, F	S, R, F	B, L, S, R, F	B, L, S, R, F	B, L, S, R, F	B, L, S, R, F
14	Syringic acid			F, L			
15	Vanillic acid	B, L, S, R, F	B, L, S, R, F	S, R, B, F, L	B, F, L	B, L, S, R, F	B, L, S, R, F

(Continued)

TABLE 1.2 (Continued)   Phytoconstituents of *Terminalia* species*

SL. NO.	COMPOUND	T. ARJUNA	T. BELLIRICA	T. CATAPPA	T. CHEBULA	T. ELLIPTICA/T. ALATA	T. PANICULATA
Flavonoids							
16	Amentoflavone		L, S		L, R, F	L	
17	Apigenin	B, L, S, R, F	B, L, S, R, F	B, L, S, R, F	B, L, S, R, F	B, L, S, R, F	B, L, S, R, F
18	Apigenin-6-C-glucoside					L	
19	Apigenin-6-C-(2″-O-galloyl)-β-D-glucoside			L			
20	Apigenin-8-C-(2″-O-galloyl)-β-D-glucoside			L			
21	Arjunolone (6,4-dihydroxy-7-methoxy flavone)	SB, B					
22	Arjunone	B, F					
23	Bicalein (5,6,7-trihydroxy flavone)	SB, B					
24	Catechin	B, L, S, R, F	B, L, S, R, F	B, L, S, R, F, SB	B, L, S, R, F	R	R, B
25	Cerasidin	F					
26	Chrysoeriol			B			
27	Cyanidin	B					
28	Cyanidin-3-glucoside			F			
29	3,4-Dimethoxy quercetin		F		F		
30	(–)-Epicatechin	B, L		B, L			B

(Continued)

TABLE 1.2 (Continued) Phytoconstituents of *Terminalia* species*

SL. NO.	COMPOUND	T. ARJUNA	T. BELLIRICA	T. CATAPPA	T. CHEBULA	T. ELLIPTICA/T. ALATA	T. PANICULATA
31	(–)-Epicatechin-3-O-gallate			B			
32	(–)-Epigallocatechin	B		B, L			
33	(–)-Epigallocatechin-3-O-gallate			B			
34	Eriodictyol	B, L, S, R, F	B, L, S, R, F	B, L, S, R, F	B, L, S, R, F	B, L, S, R, F	L, S, R, F
35	Flavanomarein			F			
36	Genistein	B, L, S, R, F	B, L, S, R, F	B, L, S, R, F	B, L, S, R, F	B, L, S, R, F	B, L, S, R, F
37	Gallocatechin	SB		B, L			
38	2″-O-Galloylisovitexin			L			
39	2″-O-Galloylvitexin			L			
40	2-O-β-Glucosyloxy-4,6,2′,4′-tetramethoxychalcone					R	
41	7-Hydroxy-3′,4-(methylenedioxy)flavan		FR, F				
42	Isoorientin	B, L, S, R, F	B, L, S, R, F	B, L, S, R, F	B, L, S, R, F	L	L
43	Isorhamnetin			B, F, W			
44	Isorhamnetin-3-O-glucoside					L	
45	Isorhamnetin-3-O-rhamnosylglucoside					L	

(Continued)

TABLE 1.2 (Continued)   Phytoconstituents of *Terminalia* species*

SL. NO.	COMPOUND	T. ARJUNA	T. BELLIRICA	T. CATAPPA	T. CHEBULA	T. ELLIPTICA/T. ALATA	T. PANICULATA
46	Isorhamnetin-3-glucoside-4'-glucoside			F			
47	Isorhamnetin pentoside		L				
48	Isovitexin	L, F		L	L, R, F	L, S, F	S, F
49	Kaempferol	B, L, S, R, F	B, L, S, F	B, L, S, R, F, SB	B, L, S, R, F	L, S, R, F	L, S, R, F
50	Kaempferol-3-O-α-L-arabinoside			WP			
51	Kaempferol galloyl-hexoside		L				
52	Kaempferol-3-O-β-D-glucopyranoside		L				
53	Kaempferol-3-O-β-D-rutinoside	B, L, S, R, F	B, L, S, R, F	L, S, F	B, L, S, R, F	L, S, R	L, R, F
54	Leucoanthocyanidin					R	
55	Leucocyanidin	B, RB					
56	Leucodelphinidin	B					
57	Luteolin	B, L,AP, S, B	L, S	L	R, L, F	L, S	L, S, F
58	Luteolin-7-O-glucoside					L	
59	8-Methyl-5,7,2',4'-tetramethoxy-flavanone					R	B
60	Morin			B			
61	Myricetin					L	

(Continued)

TABLE 1.2 (Continued)   Phytoconstituents of *Terminalia* species*

SL. NO.	COMPOUND	T. ARJUNA	T. BELLIRICA	T. CATAPPA	T. CHEBULA	T. ELLIPTICA/T. ALATA	T. PANICULATA
62	Myricetin glucoside			SB			
63	Myricetin hexoside					L	
64	Naringin	L, S, F	B, F		L, R, F	B, L, S, R	R, F
65	Orientin	L, F	B, S, L	L, R, F	B, L, S, R, F	L, S, F, R	B, S, F
66	Pelargonidin	B			F		
67	Quercetin	B, L, R, F	R, B, F, L	L, S, F, SB	L, S, R, F	F, L, AP	L, B
68	Quercetin coumaroyl-glucoside		L				
69	Quercetin-3,4'-di-O-glucoside	B, L, S, F	B, S, F	B, L, S, R, F	B, L, S, R, F	B, L, S, R, F	B, L, S, R, F
70	Quercetin-3-O-β-D-glucoside		L	L		L	
71	Quercetin-3-Glucuronide			F			
72	Quercetin-3-O-rhamnoside		L	L			
73	Quercetin-7-O-rhamnoside	F					
74	Quercetin-3-O-rutinoside		L			L	
75	Quercetin-3-D-xyloside						
76	Rutin	B, L, S, F	L, S	B, L, S, F, SB	B, L, S, R, F	L, S, F	B, L, S, R, F
77	Scutellarein	B, R	B, L, S, R, F	L, F	B, L, S, R, F	B, L, S, R, F	B, L, R, F

(Continued)

TABLE 1.2 (Continued) Phytoconstituents of *Terminalia* species*

SL. NO.	COMPOUND	T. ARJUNA	T. BELLIRICA	T. CATAPPA	T. CHEBULA	T. ELLIPTICA/T. ALATA	T. PANICULATA
78	2',4',5,7, Tetramethoxy-8-methyl flavanone						B
79	5,7,2'-Tri-O-methylflavanone4'-O-α-L-rhamnosyl-(1 → 4)-β-D-glucoside					R	
80	Vitexin	B, L, S, R, F	B, L, S, R	L, S, R, F	B, L, S, R, F	B, L, S, R, F	B, L, S, F
Lignans							
81	Anolignan B		F, R				
82	Termilignan		F, R				B
83	Thannilignan		F, R				
Phenols and glycosides							
84	Brevifolin carboxylic acid		L	F	F		
85	Chebulic acid		F, L	F	F		
86	5-Dehydroshikimic acid		F		F		
87	Digallic acid				F		
88	4-O-(2'', 4''-di-O-galloyl-α-L-rhamnosyl) ellagic acid				F		
89	4-O-(3'', 4''-di-O-galloyl-α-L-rhamnosyl) ellagic acid		F	L	F		
90	3,3'-Di-O-methylellagic acid		F	B	F		H,B

(Continued)

TABLE 1.2 (Continued) Phytoconstituents of *Terminalia* species*

SL. NO.	COMPOUND	T. ARJUNA	T. BELLIRICA	T. CATAPPA	T. CHEBULA	T. ELLIPTICA/T. ALATA	T. PANICULATA
91	3,3'-Di-O-methylellagic acid 4-mono glucoside						H
92	3,3'-Di-O-methylellagic acid 4-O-β-D-glucosyl-(1 → 4)-β-D-glucosyl-(1 → 2)-α-L-arabinoside					R	
93	3,3'-Di-O-methyl-4-O-(n"-O-galloyl-β-D-xylopyranosyl) ellagic acid. (n=2, 3,or 4)		F		F		
94	3,3'-Di-O-methyl-4-O-(β-D-xylopyranosyl) ellagic acid		F		F		
95	Ellagic acid	B, L, S, F, RB	B, L, S, R, F	SB, L, R, F	L, SB, R F	B, L, S, R, F, AP	B, L, S, R, F, H
96	Ellagic acid deoxyhexose			SB			
97	Ellagic acid 4-O-xylopyranoside			SB			
98	Eschweilenol C (ellagic acid 4-O-α-L-rhamnoside)				F		
99	Ethyl gallate	B, AP	F	SB	AP, F		
100	Gallic acid	B, L, S, R, F, RB	B, L, S, R, F	SB, F, R, L	SB, F, R, L	B, L, S, R, F, AP	B, L, S, R, F
101	Gallagic acid		L, B, F	SB	F		

(Continued)

TABLE 1.2 (Continued)   Phytoconstituents of *Terminalia* species*

SL. NO.	COMPOUND	T. ARJUNA	T. BELLIRICA	T. CATAPPA	T. CHEBULA	T. ELLIPTICA/T. ALATA	T. PANICULATA
102	4-O-(4″-O-galloyl-α-L-rhamnopyranosyl) ellagic acid		F		F		
103	4-O-galloyl-(−)-shikimic acid				F		
104	5-O-galloyl-(−)-shikimic acid				F		
105	6′-O-methyl chebulate				F		
106	7′-O-methyl chebulate				F		
107	11-Methyl chebulate				F		
108	13-Methyl chebulate				F		
109	Methyl ellagic acid	B					
110	3-O-methyl ellagic acid		F		F		
111	Methyl gallate		F, L		F		
112	3′-O-methyl-4-O-(3″,4″-di-O-galloyl-α-L-rhamnopyranosyl) ellagic acid		F		F		
113	3′-O-methyl-4-O-(n″-O-galloyl-β-D-xylopyranosyl) ellagic acid (n=2, 3, or 4)				F		
114	3-O-methyl ellagic acid 3-O-rhamnoside	B					
115	3-O-methyl ellagic acid 4′-O-α-L-rhamnopyranoside	B					
116	4-O-methylgallic acid				F		

(Continued)

TABLE 1.2 (Continued) Phytoconstituents of *Terminalia* species*

SL. NO.	COMPOUND	T. ARJUNA	T. BELLIRICA	T. CATAPPA	T. CHEBULA	T. ELLIPTICA/T. ALATA	T. PANICULATA
117	3'-O-methyl-4-O-(α-L-rhamnopyranosyl) ellagic acid	B					
118	Methyl shikimate				F		
119	3'-O-methyl-4-O-(β-D-xylopyranosyl) ellagic acid		F		F		
120	Phenol				Se		
121	O-Pentamethyl flavellagic acid						H
122	Phloroglucinol				Se		
123	Phyllemblin (ethyl gallate isomers1 progallin A)	B	F		Se		
124	Pyrogallol				Se		
125	Resveratrol			SB		B	
126	Shikimic acid		L, F	F	F		
127	Terflavin D				L		
128	3,4,5-tri-O-galloyl-shikimic acid		F		F		
129	Tetra-O-methyl ellagic acid						H
130	3,4,4'-tri-O-methyl ellagic acid			B, F			
131	3,4,3'-Tri-O-methylflavellagic acid						H
	Steroids / cardenolides						
132	Cannogenol 3-O-β-D-galactosyl-(1 → 4)-O-α-L-rhamno-side		Se	F, S			

(Continued)

TABLE 1.2 (Continued) Phytoconstituents of *Terminalia* species*

SL. NO.	COMPOUND	T. ARJUNA	T. BELLIRICA	T. CATAPPA	T. CHEBULA	T. ELLIPTICA/T. ALATA	T. PANICULATA
133	Daucosterol	B,L	F	SB, F, L	F		
134	14, 16 dianhydrogitoxigenin-3-β-D-xylopyranosyl (1→2)-O-β-D-galactopyranoside (cardenolide)	S, L					
135	16,17-Dihydroneridienone 3O-β-D-glucosyl-(1 → 6)-O-β-D-galactoside	R or RB					
136	β-Sitosterol	B,S, F, RB	F	B, SB, L	B,F	H	B,H
137	β-Sitosteryl palmitate			SB, H			
138	Stigmasterol	L				H	
139	Stigmasterol 3-O-β-D-glucoside	F					
Tannins							
140	Acutissimin A			B			
141	Arjunin	B, L			F		
142	2,3:4,6-bis-O-HHDP-1-O-galloyl-β-D-glucose	SB, L					
143	bis-HHDP-hexoside		L				
144	Castalagin	B,L		B	F		
145	Castalin			B			
146	Castamollinin			B			
147	Casuariin	B					

(Continued)

TABLE 1.2 (Continued) Phytoconstituents of *Terminalia* species*

SL. NO.	COMPOUND	T. ARJUNA	T. BELLIRICA	T. CATAPPA	T. CHEBULA	T. ELLIPTICA/T. ALATA	T. PANICULATA
148	Casuarinin	L, B		B	F		
149	Catappanin A			B			
150	Chebulagic acid	F,B, L,S,R	F, B, L,S,R	F, B, L, S, R	F, B, L, S, R, Se	F, B, L, R	F, B, L, S, R
151	Chebulaginic acid		F		F		
152	Chebulanin		F	L	F		
153	Chebulinic acid	F,B, L, S, R	F, B, L, S, R	F, B, L, S, R	F,B, L, S, R	F, B, L, S, R	F, B, L, S, R
154	2,4-Chebuloyl-β-D-glucopyranoside				F		
155	Chebumeinin A				F		
156	Chebumeinin B				F		
157	Chebupentol		F		F		
158	2-O-Cinnamoyl-1,6-di-O-galloyl-β-D-glucose				F		
159	Corilagin		F, L	L, B, F	F		
160	1-Desgalloyleugeniin			L			
161	Digalloyl glucose		L				
162	1,3-Di-O-galloyl-2,4-chebuloyl-β-D-glucose				F		
163	1,6-Di-O-galloyl-2,4-chebuloyl-β-D-glucose		F		F		

(Continued)

TABLE 1.2 (Continued) Phytoconstituents of *Terminalia* species*

SL. NO.	COMPOUND	T. ARJUNA	T. BELLIRICA	T. CATAPPA	T. CHEBULA	T. ELLIPTICA/T. ALATA	T. PANICULATA
164	1,6-Di-O-galloyl-2-O-cinnamoyl-β-D-glucose				F		
165	1,2-Di-O-galloyl-6-O-cinnamoyl-β-D-glucose				F		
166	3,6-Di-O-galloyl-D-glucose				F		
167	1,6-Di-O-galloyl-β-D-glucose		L, F		F		
168	1,3-Di-O-galloyl-β-D-glucose				F		
169	Digalloyl-HHDP-hexoside		L				
170	3,5-Di-methoxy-4-hydroxyphenol-1-O-β-D-(6'-O-galloyl)-glucoside			B			
171	Dimethyl 4'-epi-neochebulagate				F		
172	Dimethyl neochebulagate				F		
173	Dimethyl neochebulinate				F		
174	Ellagic tannin		L				
175	Eugenigrandin A			B			
176	epi-Catechin-(epi)catechin		L				
177	epi-Catechin-(epi)gallocatechin		L				
178	Ethyl flavogallonate			SB			
179	Flavogallonic acid		F	SB	F		
180	Epi-gallocatechin-gallate			SB			

(Continued)

TABLE 1.2 (Continued) Phytoconstituents of *Terminalia* species*

SL. NO.	COMPOUND	T. ARJUNA	T. BELLIRICA	T. CATAPPA	T. CHEBULA	T. ELLIPTICA/T. ALATA	T. PANICULATA
181	6-O-Gallotannic (tannic acid)				F		
182	3-O-galloyl-epicatechin			B			
183	3-O-galloyl-epigallocatechin			B,L			
184	Galloyl-D-glucose	L	F, L		F		
185	Galloyl-HHDP-hexoside		L				
186	3′-O-galloyl procyanidin B-2			B			
187	Galloylpunicalagin		L				
188	2-O-Galloylpunicalin	B					
189	Gemin D				F		
190	Geranin			L	F		
191	Granatin B		L	L			
192	Grandinin			B			
193	2,3-(S)-HHDP-6-O-galloyl-D-glucose	B					
194	2,3-O-(S)-HHDP-D-glucose	B	L	B, L			
195	2,3:4,6-bis-O-HHDP-1-O-galloyl-β-D-glucose	B,L					
196	HHDP methyl ester		F				
197	3-Methoxy-4-hydroxyphenol-1-O-β-D-(6′-O-galloyl)-glucoside			B			
198	3′-Methoxy quercetin				F		

(Continued)

TABLE 1.2 (Continued) Phytoconstituents of *Terminalia* species*

SL. NO.	COMPOUND	T. ARJUNA	T. BELLIRICA	T. CATAPPA	T. CHEBULA	T. ELLIPTICA/T. ALATA	T. PANICULATA
199	Methyl chebulagate				F		
200	Methylflavogallonate		F		F		
201	6'-O-methyl neochebulagate		F		F		
202	1'-O-methyl neochebulanin				F		
203	1'-O-methyl neochebulinate 1 or 6		F		F		
204	Neochebulagic acid				F		
205	Neochebulinic acid				F		
206	Pedunculagin				F		
207	1,2,3,4,6-penta-O-galloyl-β-D-glucose	L	F		L, F		
208	Phyllanemblinin E				F		
209	Phyllanemblinin F				F		
210	Procyanidin B-1			B		B	
211	Procyanidin B-2					B	
212	Procyanidin B-3					B	
213	Punicacortein A		L		F		
214	Punicacortein C				F		
215	Punicacortein D				F		
216	Punicafolin		L				
217	α/β-Punicalagin -	B	F, L	B,L	L, F		

(Continued)

TABLE 1.2 (Continued) Phytoconstituents of *Terminalia* species*

SL. NO.	COMPOUND	T. ARJUNA	T. BELLIRICA	T. CATAPPA	T. CHEBULA	T. ELLIPTICA/T. ALATA	T. PANICULATA
218	Punicalin	B,L	F	B,L	L, F		
219	Puniglucolin		L				
220	Tellimagrandin I		F	B,L, SB	F		
221	Tellimagrandin II			B,L, SB	F		
222	Tercatain			B, L,F	F		
223	Terchebulin	B		SB	F		
224	Terflavin A			L, SB	F		
225	Terflavin B		F,L	L	L, F		
226	Terflavin C	B		L	L,F		
227	Terflavin D				F, L		
228	Tergallagin			L			
229	1,2,3,6-Tetra-O-galloyl-4-O-cinnamoyl-β-D-glucose				F		
230	1,2,3,4-Tetra-O-galloyl-β-D-glucose				F		
231	1,2,3,6-Tetra-O-galloyl-β-D-glucose		F		F		
232	1,3,4,6-tetra-O-galloyl-β-D-glucose		F		F		
233	2,3,4,6-Tetra-O-galloyl-D-glucose	SB, L			L		
234	1,2,3-Tri-O-galloyl-6-O-cinnamoyl-β-D-glucose				F		
235	1,2,6-Tri-O-galloyl-β-D-glucopyranose				F		

(Continued)

TABLE 1.2 (Continued) Phytoconstituents of *Terminalia* species*

SL. NO.	COMPOUND	T. ARJUNA	T. BELLIRICA	T. CATAPPA	T. CHEBULA	T. ELLIPTICA/T. ALATA	T. PANICULATA
236	1,3,6-Tri-O-galloyl-β-D-glucose (terchebin)			B, F	F, G		
237	2,3,6-Tri-O-galloyl-β-D-glucose		F		F		
238	3,4,6-tri-O-galloyl-β-D-glucose		F, L		F		
239	1,3,4,6-Tetra-O-galloyl-2-O-cinnamoyl-β-D-glucose				F		
240	Triethyl chebulate				F		
241	2,3,6-tri-O-galloyl-beta-D-glucose		F		F		
Triterpenoids							
242	3-Acetylmaslinic acid					RB	
243	β-Amyrin					R	
244	Arjuna homosesquiterpenol					B	
245	Arjunaside A	B					
246	Arjunaside B	B					
247	Arjunaside C	B					
248	Arjunaside D	B					
249	Arjunaside E	B					
250	Arjunetin	B, L, S, R, F	B, L, S, R, F	B, L, S, R, F	B, L, S, R, F	B, L, S, R, F	B, L, S, R, F
251	Arjunetoside	R, SB					

(Continued)

TABLE 1.2 (Continued) Phytoconstituents of *Terminalia* species*

SL. NO.	COMPOUND	T. ARJUNA	T. BELLIRICA	T. CATAPPA	T. CHEBULA	T. ELLIPTICA/T. ALATA	T. PANICULATA
252	Arjungenin	SB, L, R, B	B, L, S, R, F	B, L, S, R, F	B, L, S, R, F	B, L, S, R, F	B, L, S, R, F
253	Arjunglucoside I	B, R	B, F		B, F		
254	Arjunglucoside II	B		SB	F		
255	Arjunglucoside III	B					
256	Arjunglucoside IV	B					
257	Arjunglucoside V	B					
258	Arjunic acid	SB, F, B,R			F	B	
259	Arjunolic acid	B, H, L, S, R, F, RB	B, L, S, R	B, L, S, R, F, SB	B, L, S, R, F	B, H, L, S, R, F	B, L, S, R, F
260	Arjunolitin	SB, B					
261	Arjunoside I	SB, RB, B					
262	Arjunoside II	SB, RB, B					
263	Arjunoside III	R, RB					
264	Arjunoside IV	R, RB					
265	Asiatic acid			L			
266	Barringtogenol					H, B	
267	Belleric acid		B, F				
268	Bellericagenin A		B				
269	Bellericagenin B		B				

(Continued)

TABLE 1.2 (Continued) Phytoconstituents of *Terminalia* species*

SL. NO.	COMPOUND	T. ARJUNA	T. BELLIRICA	T. CATAPPA	T. CHEBULA	T. ELLIPTICA/T. ALATA	T. PANICULATA
270	Bellericanin		F				
271	Bellericanin glycoside		F				
272	Bellericaside A		B				
273	Bellericaside B		B				
274	Bellericoside	B	B, F		B		
275	Betulinic acid	B, L, S, R, F	SB	B, L, S, R, F, SB	B, L, S, R, F		B, L, S, R
276	Chebuloside I				F		
277	Chebuloside II			SB	F		
278	Crataegioside	B			F		
279	2α,19α-Dihydroxy-3-oxo-olean-12-en-28-oic acid-28-O-β-D-glucoside	R					
280	2α,3β-Dihydroxyurs-12,18-dien-28-oic acid-28-O-β-D-glucopyranoside	B					
281	2α,19α-Dihydroxy-olean-12-en-28-methylester-3β-O-rutinoside					R	
282	Friedelin	F	S			R	
283	23-O-galloylarjunic acid				F		B
284	23-O-galloylarjunolic acid 28-O-β-D-glucosyl ester				F		

(Continued)

TABLE 1.2 (Continued) Phytoconstituents of *Terminalia* species*

SL. NO.	COMPOUND	T. ARJUNA	T. BELLIRICA	T. CATAPPA	T. CHEBULA	T. ELLIPTICA/T. ALATA	T. PANICULATA
285	23-O-galloylarjunolic acid				F		
286	23-O-galloylpinfaenoic acid 28-O-β-D-glucosyl ester				F		
287	23-O-galloylterminolic acid 28-O-β-D-glucosyl ester				F		
288	19α-Hydroxy asiatic acid	F					
289	2α-Hydroxymicromeric acid				F	L	
290	2α-Hydroxyursolic acid				F	L	
291	Kajiichigoside F1	B, F				L	
292	Lupeol	SB					
293	Maslinic acid				F	L, H	
294	Maslinic lactone					H	
295	Methyl oleanate	R, F				R	
296	23-O-4'-epi-Neochebuloylarjungenin				F		
297	23-O-neochebuloylarjungenin 28-O-β-D-glycosyl ester				F		
298	Olean-12-en-2α,3β-diol	F					
299	Olean-3α,22β-diol-12 en-28-oic acid 3-O-β-D-glucosyl-(1 → 4)-β-D-glucoside	B					

(Continued)

TABLE 1.2 (Continued) Phytoconstituents of *Terminalia* species*

SL. NO.	COMPOUND	T. ARJUNA	T. BELLIRICA	T. CATAPPA	T. CHEBULA	T. ELLIPTICA/T. ALATA	T. PANICULATA
300	Olean 3β,6β,22α-triol-12en-28-oic acid-3-O-β-D-glucosyl-(1 → 4)-β-D-glucoside	B					
301	Oleanolic acid	B, L, S, R, F, RB	B, L, S, R	B, L, S, R, F	B, L, S, R	B, L, S, R, F	L, S, R, F
302	Pinfaenoic acid 28-O-β-D-glucosyl ester	B			F		
303	Quadranoside I	B					
304	Quadranoside VIII	B					
305	Quercotriterpenoside I				F		
306	Sericoside	B					
307	Termiarjunoside I (olean-1α,3β,9α,22α-tetraol-12-en-28-oic acid-3-β-D-glucoside)	SB, B					
308	Termiarjunoside II (olean-3α,5α,25-triol-12-en-23,28-dioic acid-3α-D-glucoside)	SB, B					
309	Terminic acid	R, H, B, RB					
310	Terminolic acid	B		H	F, B	H	

(Continued)

TABLE 1.2 (*Continued*) Phytoconstituents of *Terminalia* species*

SL. NO.	COMPOUND	T. ARJUNA	T. BELLIRICA	T. CATAPPA	T. CHEBULA	T. ELLIPTICA/T. ALATA	T. PANICULATA
311	Terminolitin	B,F					
312	Terminoside A	B					
313	2α,3β,19β,23-Tetrahydroxyolean-12-en-28-oic acid 3β-O-β-D-galactosyl-(1 → 3)-β-D-glucoside-28-O-β-D-glucoside					R	
314	Tormentic acid	B, F				H	
315	2α,3β,19α-Trihydroxyolean-12-en-28-oic acid 3-O-β-D-galactosyl-(1 → 3)-β-D-glucoside					R	
316	2α,3β,19α-Trihydroxyolean-12-en-28-oic acid methylester 3β-O-rutinoside	R				R	
317	2α,3β,23 trihydroxyurs-12,18-dien-28-oic acid-28-O-β-D-glucopyranoside	B					
318	2α,3β,23-trihydroxyurs-12-en-28-oic acid			L			
319	2α,3β,19α-trihydroxyurs-12-en (Torment)	F					
320	Ursolic Acid	B, L, S, R, F	L, S, R	B, L, S, R, F	L, S, R	B, L, S, R, F	B, L, S, F

(*Continued*)

TABLE 1.2 (Continued)   Phytoconstituents of *Terminalia* species*

SL. NO.	COMPOUND	T. ARJUNA	T. BELLIRICA	T. CATAPPA	T. CHEBULA	T. ELLIPTICA/T. ALATA	T. PANICULATA
Others							
321	Ascorbic acid			F, Se	F		
322	Anthraquinone glycoside		F		F		
323	Arachidic acid		Se	Se	F, Se		
324	Arachidic stearate	F					
325	Arjunaphthanoloside	SB					
326	Behenic acid		B, F	Se	F, Se		
327	Benzoyl-β-D-(4' → 10"geranilanoxy)-pyranoside		F				
328	β-Carotene			L, F			
329	Citric acid		Se				
330	Dioleolinolein		F				
331	Geranyl-10-oxy-10-O-β-D-xylopyranoside 2'-benzoate		AP				
332	Hentriacontane	F	F				
333	Hydroxy anthraquinones glycoside				F		
334	8-Hydroxyl hexadecanoic acid	RB					
335	Linoleic acid		Se	F. oil	F		
336	Linolenic acid			F. oil	Se		

(Continued)

TABLE 1.2 (Continued) Phytoconstituents of *Terminalia* species*

SL. NO.	COMPOUND	T. ARJUNA	T. BELLIRICA	T. CATAPPA	T. CHEBULA	T. ELLIPTICA/T. ALATA	T. PANICULATA
337	Mangiferin	B, S, F	B, R, F	L, R, F	B, L, S, R, F	L, S, R	B, L, S, F
338	Myristic acid			F. oil			
339	Myristyl oleate	F					
340	3,4,8,9,10-Pentahydroxydibenzo[b,d]pyran-6-one		F				
341	Oleic acid		Se	F. oil	F, Se		
342	Palmitic acid		Se	F, F. oil	FR, Se		
343	Palmitoleic acid			F. oil			
344	2-pentadecanone			F			
345	Ricinoleic acid				Se		
346	Stearic acid			F, F.oil	F, Se		
347	3,4,5-Trimethoxyphenyl-1-O-(4-sulfo)-β-D-glucopyranoside (B)			SB			

* R root, SB stem bark, B bark, F fruit, S stem, H heartwood, RB root bark, Se seed, FR fruit rind, WP whole plant, G gall, AP aerial part, F.oil fruit oil

terflavin C, terchebulin, casurarinin, casuariin and castalagin. Ellagic acid and its glycoside 3′-O-methyl-4-O-(α-L-rhamnopyranosyl) ellagic acid are also reported from *T. arjuna*. *T. arjuna* is also a good source of flavonoids.

Flavonols quercetin and kaempferol, flavones luteolin, apigenin, arjunolone, baicalein, vitexin, isovitexin and arjunone, flavan-3-ols catechin, epicatechin, gallocatechin and epigallocatechin, anthocyanidin, pelargonidin and leucocyanidin are the flavonoids found in *T. arjuna* (Saha et al. 2012; Fahmy et al. 2015; Amalraj and Gopi 2017). From the ethyl acetate fraction of the alcoholic extract of the seeds of *T. arjuna*, a new cardenolide 14,16-dianhydrogitoxigenin-3-O-β-D-xylopyranosyl (1 → 2) -O-β-D-galactopyranoside was isolated (Yadava and Rathore 2000).

The main constituents of *T. bellirica* are lignans, tannins, terpenoids, flavonoids, phenolic acids, phenols, glycosides and sugars. Gallic acid and gallate esters reported in *T. bellirica* include gallic acid, methyl gallate, 1,6-di-O-galloyl-β-D-Glc, 3,4,6-tri-O-galloyl-β-D-Glc, 1,3,4,6-tetra-O-galloyl-β-D-Glc, 1,2,3,4,6-penta-O-galloyl-β-D-Glc and 3,4,5-tri-O-galloyl-shikimic acid. Chebulic acid and chebulic ellagitannins reported are chebulic acid, chebulanin, chebulinic acid, methyl neo-chebulanin, methyl neochebulinate, chebulagic acid, methyl neochebulagate and 1,6-di-O-galloyl-2,4-O-chebuloyl-β-D-Glc. Tellimagrandin I, corilagin, punicalin, punicalagin and terflavin B are the nonchebulic ellagitannins found in *T. bellirica*. Ellagic acid, 3-O-methyl ellagic acid, 3,3′-di-O-methyl ellagic acid, 3,4,8,9,10-pentahydroxydibenzo[b, d]pyran-6-one, flavogallonic acid, methylflavogallonate and gallagic acid are the ellagic acid derivatives reported. Ellagic acid glycosides reported are 3′-O-methyl-4-O-(β-D-xylopyranosyl) ellagic acid, 3,3′-di-O-methyl-4-O-(β-D-xylopyranosyl) ellagic acid, 3′-O-methyl-4-O-(n″-O-galloyl-β-D-xylopyranosyl) ellagic acid, 3,3′-di-O-methyl-4-O-(n″-O-galloyl-β-D-xylopyranosyl) ellagic acid, 4-O-(4″-O-galloyl-α-L-rhamnopyranosyl)ellagic acid, 4-O-(3″, 4″-di-O-galloyl-α-L-rhamnopyranosyl) ellagic acid and 3′-O-methyl-4-O-(3″, 4″-di-O-galloyl-α-L-rhamnopyranosyl) ellagic acid. Flavonoids quercetin and 7-hydroxy-3′, 4′-methylenedioxyflavan are reported. Triterpene arjungenin, belleric acid, belliricagenin A and belliricagenin B are the triterpenoids found in *T. bellirica* (Mahato et al. 1992). Phytosterols β-sitosterol and daucosterol are also found in *T. bellirica*. Arjunglucoside I, belliricaside A, belliricaside B and bellericoside are the triterpene glycosides reported (Fahmy et al. 2015; Nandy et al. 1989). Cardenolide, cannogenol 3-O-β-D-galactopyranosyl-(1→4)-O-α-L-rhamnopyranoside, was isolated from the seeds of *T. bellirica* (Yadava and Rathore 2001). *T. bellirica* also contains lignans termilignan, thannilignan and anolignan B (Valsaraj et al. 1997). Fifty compounds, mainly ellagitannins and proanthocyanidins, were tentatively identified by LC-MS in methanol extracts of leaves of *T. bellirica* (Sobeh et al. 2019). Geranyl-10-oxy-10-O-β-D-xylopyranoside 2′-benzoate was isolated from aerial parts of *T. bellirica* (Sultana et al. 2018).

Flavonoids, hydrolyzable ellagitannins and other tannin-related compounds have been isolated from the leaves, bark, seeds and fruits of *T. catappa*. Tannins were the most abundant compounds in the hydroalcoholic extract of leaves of *T. catappa* and the major compounds correspond to α and ß punicalagins (Mininel et al. 2014). Under the gallic acid and simple gallate esters, only gallic acid is reported in *T. catappa*. Non-chebulagic ellagitannins corilagin, tercatain, punicalin, punicalagin, tergallagin, terfavin A, terfavin B and terfalvain C are found in the leaves of *T. catappa* (Fahmy et al. 2015). Fruits/leaves also contain ellagic acid, gallagic acid and 3,4,4′-tri-O-methyl ellagic acid (Mininel et al. 2014). Ferulic acid, vanillic acid, coumaric acid and p-hydroxybenzoic acid are reported in leaves/fruits. Flavonoids orientin, isoorientin, vitexin and isovitexin, 2″-O-galloylvitexin, 2″-O-galloylisovitexin, gallocatechin, epicatechin, 3-O-galloyl-epicatechin, epigallocatechin and 3-O-galloyl-epigallocatechin are reported from stem bark/leaves. Ursolic acid, 2α, 3β, 23-trihydroxyurs-12-en-28-oic acid, terminolic acid, daucosterol and β-sitosterol are the only triterpenes and phytosterols found in heartwood/leaves of *T. catappa* (Fahmy et al. 2015).

From the barks of *T. catappa*, a new aromatic compound 3,4,5-trimethoxyphenyl-1-O-(4-sulfo)-β-D-glucopyranoside, two triterpenoid saponins (chebuloside II and arjunoglucoside II), two triterpenes (arjunolic acid and 3-betulinic acid) and sitosterol-3-O-β-D-glucopyranoside have been isolated (Monnet et al. 2012). Two new flavone glycosides with galloyl substitution apigenin 6-C-(2″-O-galloyl)-β-D-glucopyranoside and apigenin 8-C-(2″-O-galloyl)-β-D-glucopyranoside and four known flavone glycosides, isovitexin, vitexin, isoorientin and rutin, were also found in the ethanol extract of the dried fallen leaves of *T. catappa* L. (Lin et al. 2000).

T. chebula is a 'wonder herb' as it is a storehouse of tannins. The major phytocomponents in *T. chebula* are hydrolyzable tannins like chebulinic acid, gallotannins, phenolics like gallic acid, ellagic acid and chebulic acid, anthraquinones, triterpenoids and other miscellaneous compounds (Walia and Arora 2013). So far, non-hydrolyzable tannins are not reported in *T. chebula*. Tannins are the most important components in *T. chebula* as they account for the various medicinal properties. The dried fruit of *T. chebula* is mostly used for medicinal purposes. The average tannin content in the dried fruit pulp and dried pericarp of the seed is 30%–35% and depends on the place of origin (Khare 2004). Gallic acid, methyl gallate, ethyl gallate, galloyl glucosides 1,6-di-O-galloyl-β-D-Glc, 3,4,6-tri-O-galloyl-β-D-Glc, 1,3,4,6-tetra-O-galloyl-β-D-Glc, 2,3,4,6-tetra-O-galloyl-β-D-Glc and 1,2,3,4,6-penta-O-galloyl-β-D-Glc and 3,4,5-tri-O-galloyl-shikimic acid are reported from *T. chebula*. Chebulic acid, neochebulic acid, chebulanin, chebulinic acid, methyl neochebulanin, methyl neochebulinate, chebulagic acid, methyl neochebulagate and 1,6-di-O-galloyl-2,4-O-chebuloyl-β-D-Glc are the chebulic ellagitannins present in *T. chebula*.

The non-chebulic ellagitannins present in *T. chebula* include tellimagrandin(I), corilagin, punicalin, punicalagin, terflavin A, terflavin B, terflavin C, terchebulin and casurarinin.

Ellagic acid, 3-O-methyl ellagic acid, 3,3'-dimethyl ellagic acid, 3,4,8,9,10-pentahydroxydibenzo[b,d]pyran-6-one, flavogallonic acid, methyl flavogallonate and gallagic acid are the ellagic derivatives found in *T. chebula*. The ellagic acid glycosides present are 3'-O-methyl-4-O-(β-D-xylopyranosyl)ellagic acid, 3,3'-di-O-methyl-4-O-(β-D-xylopyranosyl) ellagic acid, 3'-O-methyl-4-O-(n''-O-galloyl-β-D-xylopyranosyl) ellagic acid (n=2, 3,or 4), 3,3'-di-O-methyl-4-O-(n''-O-galloyl-β-D-xylopyranosyl) ellagic acid (n=2, 3, or 4), 4-O-(4''-O-galloyl-α-L-rhamnopyranosyl) ellagic acid, 4-O-(3'', 4''-di-O-galloyl-α-Lrhamnopyranosyl) ellagic acid and 3'-O-methyl-4-O-(3'', 4''-di-O-galloyl-α-L-rhamnopyranosyl) ellagic acid. The phenolic acids reported are caffeic acid, ferulic acid, vanillic acid and coumaric acid, whereas the only flavonoids reported are flavonol quercetin, rutin and flavone luteolin. The triterpenoids found in *T. chebula* include 2α-hydroxyursolic acid, maslinic acid and 2α-hydroxymicromeric acid, whereas the phytosterols found are β-sitosterol and daucosterol. Some of the triterpene glycosides found in *T. chebula* are chebuloside I and II, arjunglucoside I and bellericoside (Fahmy et al. 2015; Bag et al. 2013). From methanolic extracts of the dried fruits of *T. chebula*, 48 tannins and 12 polyhydroxytriterpenoid derivatives were isolated (Kim et al. 2018).

Ellagic acid derivative 3,3'-di-O-methyl-4-O-(β-Dglucopyranosyl-(1→4)-β-D-glucopyranosyl-(1→2)-α-L-arabinopyranosyl) ellagic acid, flavonoid 8-methyl-5,7,2', 4'-tetra-O-methylflavanone, 5,7,2'-tri-O-methylflavanone 4'-O-α-L-rhamnopyranosyl-(1→4)-β-D-glucopyranoside, chalcone 2-O-β-D-glucosyloxy-4,6,2', 4'-tetramethoxy chalcone, triterpene tomentosic acid, 3-acetylmaslinic acid, friedelin, triterpene glycosides 2α, 3β, 19α-trihydroxyolean-12-en-28-oic acid-methylester-3-O-rutinoside, 2α, 3β, 19α-trihydroxyolean-12-en-28-oic acid-3-O-β-D-galactopyranosyl-(1→3)-β-D-glucopyranoside, 2α, 3β, 19β, 23-tetrahydroxyolean-12-en-28-oic acid and 3-O-β-D-galactopyranosyl-(1→3)-β-D-glucopyranoside-28-O-β-D-glucopyranoside are some of the phytoconstituents reported from *T. alata* syn. *T. elliptica* (Fahmy et al. 2015). Triterpenoids like arjunic acid, arjunolic acid, arjunetin, oleanolic acid and betulinic acid and phytosterols stigmasterol and β-sitosterol are reported from *T. alata* (Mallavarapu et al. 1980, 1986; Row and Rao 1962a; Anjaneyulu et al. 1986). Arjunolic acid, barringtogenol, oleanolic acid, tomentosic acid and β-sitosterol were isolated from the heartwood of *T. tomentosa* (Row and Rao 1962a). From ethanolic extract of stem bark of *T. arjuna*, homosesquiterpenol was isolated (Joshi et al. 2013). Quercetin-3-O-rutinoside, luteolin-7-O-glucoside, myricetin hexoside, quercetin-3-O-glucoside, isorhamnetin-3-O-rhamnosylglucoside and isorhamnetin-3-O-glucoside were identified in the methanolic extract of *T. coriacea* leaves (Khan et al. 2017). Three new gly-cosides 3,3'-di-*O*-methylellagic acid 4-*O*-β-D-glucopyranosyl-(1→4)-β-D-glucopyranosyl-(1→2)-α-L-arabinopyranoside, 5,7,2'-tri-*O*-methylflavanone 4'-*O*-α-L-rhamnopyranosyl-(1→4)-β-D-glucopyranoside and 2α, 3β, 19β, 23-tetrahydroxyolean-12-en-28-oic acid 3-*O*-β-D-galactopyranosyl-(1→3)-β-D-glucopyranoside-28-*O*-β-D-glucopyranoside were isolated from the roots of *T. alata* (Srivastava et al. 2001). Procyanidin B1, B2 and B3 and resveratrol were identified by LC-MS (Gahlaut et al. 2013). LC-MS analysis also identified quercetin-3-O-rutinoside, luteolin-7-O-glucoside, myricetin hexoside, quercetin-3-O-glucoside, isorhamnetin-3-O-rhamnosylglucoside and isorhamnetin-3-O-glucoside in the leaf methanolic extract of *T. coriacea* (Khan et al. 2018).

Phytochemicals such as ellagic acid, 3, 3'-O-dimethylellagic acid, 3, 3'-O-dimethylellagic acid-4-glucoside, O-pentamethyl flavellagic acid, 3, 4, 3'-O-trimethyl flavellagic acid and β-sitosterol have been isolated from the heartwood of *T. paniculata* (Row and Rao 1962b; Row and Raju 1967). The polyphenol-rich fraction of bark of *T. paniculata* was found to contain gallic acid, ellagic acid and quercetin (Ganjayi et al. 2017). An aqueous extract of the bark of *T. paniculata* contained gallic acid, ellagic acid, catechin and epicat-echin (Ramachandran et al. 2013).

1.4 PHARMACOLOGICAL ACTIVITIES

Polyphenols and triterpenoids are the most common class of compounds in the *Terminalia* genus. The predominant polyphenols are hydrolyzable tannins (gallotannins and ellagitannins). Oleanoic acid derivatives are the major con-stituents of the triterpenoids class of compounds. Besides these, phenolic acids and fatty acids are also present. The presence of such potent molecules renders the genus with varied biological activities including antidiabetic, antihyperlip-idemic, antioxidant, antibacterial, antifungal, antiviral, antiinflammatory, anti-cancer, antiulcer, antiparasitic, cardioprotective, hepatoprotective, hypotensive, hypolipidemic, gastroprotective and wound healing activities (Table 1.3). There are several reviews on the biological activities of *Terminalia* species (Fahmy et al. 2015; Upadhyay et al. 2014; Dwevedi et al. 2016; Cock 2015; Anand et al. 2015; Beigi et al. 2018). Other examples are phytopharmacological reviews of *T. arjuna* (Kumar 2014), *T. bellirica* (Deb et al. 2016; Kumari et al. 2017; Singh et al. 2018a; Arabind et al. 2019; Khaliq and Fahim 2018; Belapurkar et al. 2014), *T. catappa* (Anand et al. 2015; Venkatalakshmi et al. 2016a; Mallik et al. 2013)

TABLE 1.3 Pharmacological activities of *Terminalia* species

	ACTIVITY	SPECIES	PART USED	REFERENCE
1	Antihelmintic	*T. arjuna*	Bark	Bodke et al. 2013
		T. paniculata	Root	Acharyya et al. 2019
2	Antiaging	*T. arjuna*	Bark	Satardekar and Deodhar 2010; Farwick et al. 2013
		T. catappa	Leaf	Wen et al. 2010
		T. chebula	Fruit	Satardekar and Deodhar 2010
3	Antianthrax	*T. arjuna*	Branch	Wright et al. 2016b
		T. catappa	Fruit	Wright et al. 2016b
		T. chebula	Fruit	Wright et al. 2016b
4	Antiarthritis	*T. chebula*	Fruit	Nair et al. 2010; Seo et al. 2012
		T. tomentosa	Bark	Jitta et al. 2019
5	Antiatherosclerotic	*T. arjuna*	Bark	Shaila et al. 1998
		T. bellirica	Fruit	Shaila et al. 1995; Tanaka et al. 2016
6	Antibacterial	*T. arjuna*	Leaf, branch	Shinde et al. 2009; Wright et al. 2016a
		T. bellirica	Leaf	Shinde et al. 2009
		T. catappa	Bark, fruit, leaf	Sangavi et al. 2015; Wright et al. 2016a; Opara et al. 2012
		T. chebula	Leaf, fruit	Shinde et al. 2009; Wright et al. 2016a
		T. elliptica	Leaf	Shinde et al. 2009
7	Anticancer	*T. arjuna*	Bark	Pettit et al. 1996
		T. bellirica	Fruit	Li et al. 2018
		T. catappa	Leaf	Yang et al. 2010
		T. chebula	Fruit	Reddy et al. 2009
8	Antidiabetic	*T. arjuna*	Leaf	Anam et al. 2009
		T. bellirica	Fruit, bark	Nampoothiri et al. 2011; Nguyen et al. 2016
		T. catappa	Fruit	Nagappa et al. 2003
		T. chebula	Fruit, seed	Senthilkumar 2008; Rao and Nammi 2006

(Continued)

TABLE 1.3 (Continued)   Pharmacological activities of *Terminalia* species

	ACTIVITY	SPECIES	PART USED	REFERENCE
		T. elliptica	Bark	Nguyen et al. 2016
		T. paniculata	Bark	Ramachandran et al. 2012, 2013
9	Antiepileptic	*T. elliptica*	Leaf	Pasha et al. 2013, 2014
10	Antifungal	*T. arjuna*	Leaf	Mandoli et al. 2013a
		T. catappa	Leaf	Mandoli et al. 2013b
		T. chebula	Seed	Vonshak et al. 2003
		T. elliptica	Root	Srivastava et al. 2001
11	Antihyperglycemic	*T. chebula*	Fruit	Murali et al. 2004
		T. elliptica	Bark	Alladi et al. 2012
12	Antihyperlipidemic	*T. arjuna*	Bark	Ram et al. 1997
		T. bellirica	Fruit	Shaila et al. 1998
		T. catappa	Leaf	Sharma and Mukundan 2013
		T. chebula	Fruit	Israni et al. 2010
13	Antihypertensive	*T. arjuna*	Bark	Sandhu et al. 2010
		T. bellirica	Fruit	Khan and Gilani 2008
		T. catappa	Fruit	Adefegha et al. 2017
		T. chebula	Fruit	Sornwatana et al. 2015
14	Antiinflammatory	*T. arjuna*	Bark	Halder et al. 2009
		T. catappa	Leaf	Lin et al. 1999
		T. elliptica	Leaf	Khan et al. 2017
		T. paniculata	Bark	Talwar et al. 2011
15	Antimicrobial	*T. bellirica*	Stem, leaf, fruit	Chandra et al. 2013; Elizabeth 2005
		T. chebula	Fruit, leaf	Malekzadeh et al. 2001; Mostafa et al. 2011
16	Antimutagenic	*T. chebula*	Fruit	Gandhi and Nair 2005
17	Antinociceptive	*T. arjuna*	Bark	Halder et al. 2009
		T. bellirica	Fruit	Kaur and Jaggi 2010
		T. chebula	Fruit	Kaur and Jaggi 2010
		T. elliptica	Leaf	Khan et al. 2011
18	Antiobesity	*T. elliptica*	Bark	Meriga et al. 2017
		T. paniculata	Bark	Ganjayi et al. 2017; Srinivasan et al. 2016

(Continued)

TABLE 1.3 (Continued) Pharmacological activities of *Terminalia* species

	ACTIVITY	SPECIES	PART USED	REFERENCE
19	Antioxidant	*T. arjuna*	Bark, leaf, fruit	Bajpai et al. 2005; Chatha et al. 2014
		T. bellirica	Bark, leaf, fruit	Bajpai et al. 2005; Nguyen et al. 2016
		T. bellirica	Leaf, fruit	Pfundstein et al. 2010; Alam et al. 2011
		T. catappa	Leaf, fruit, nuts	Abdulkadir 2015; Etienne et al. 2017
		T. chebula	Bark, leaf, fruits	Bajpai et al. 2005; Cheng et al. 2003; Mahesh et al. 2009
		T. elliptica	Bark, leaf	Sharma et al. 2013; Pasha et al. 2014
		T. paniculata	Bark	Mopuri and Meriga 2014
20	Antistress	*T. chebula*	Fruit	Debnath et al. 2011
21	Antiulcerogenic	*T. chebula*	Fruit	Sharma et al. 2011
22	Antiviral	*T. arjuna*	Bark	Cheng et al. 2002
		T. chebula	Bark	Lee et al. 2011
23	Antiviral/HIV	*T. paniculata*	Leaf, fruit	Narayan and Rai 2016; Durge et al. 2017
24	Anxiolytic	*T. arjuna*	Bark	Sekhar et al. 2017
25	Cardioprotective	*T. arjuna*	Bark	Manna et al. 2007a,b
		T. catappa	Leaf	Bagalkote et al. 2018
		T. chebula	Fruit	Suchalatha and Devi 2004
26	Cytotoxic	*T. arjuna*	Bark, stem, leaf	Pettit et al. 1996
		T. catappa	Leaf	Chen et al. 2000
		T. chebula	Fruit	Saleem et al. 2002
27	Gastroprotective	*T. arjuna*	Bark	Devi et al. 2008
		T. catappa	Bark	Nunes et al. 2012
		T. chebula	Fruit	Mishra et al. 2013; Wright et al. 2017
		T. elliptica	Leaf	Khan et al. 2017
28	Hepatoprotective	*T. arjuna*	Fruit	Ghosh et al. 2010

(Continued)

TABLE 1.3 (Continued) Pharmacological activities of *Terminalia* species

	ACTIVITY	SPECIES	PART USED	REFERENCE
		T. bellirica	Leaf	Kinoshita et al. 2007
		T. catappa	Bark, leaf	Vahab and Harindran 2016; Jadon et al. 2007
		T. chebula	Fruit	Lee et al. 2007
		T. elliptica	Leaf	Patel et al. 2017
		T. paniculata	Bark	Eesha et al. 2011; Srinivasan et al. 2016
29	Hypolipidemic	*T. arjuna*	Bark	Jensi and Gopu 2018
		T. chebula	Fruit	Ahirwar et al. 2003
30	Spasmogenic	*T. chebula*	Seed	Mard et al. 2011
31	Wound healing	*T. arjuna*	Bark	Chaudhari and Mengi 2006
		T. bellirica	Fruit	Saha et al. 2011
		T. catappa	Bark	Khan et al. 2014
		T. chebula	Leaf, fruit	Seguna et al. 2002; Li et al. 2011
		T. elliptica	Bark	Khan et al. 2012

and *T. chebula* (Hedina et al. 2016; Rathinamoorthy and Thilagavathi 2014; Sawant et al. 2013; Belapurkar et al. 2014; Singh and Malhotra 2017; Ashwini et al. 2011).

No individual reviews on *T. elliptica* and *T. paniculata* are located, but they are dealt with in the general reviews on *Terminalia* (Cock 2015; Fahmy et al. 2015; Chang et al. 2019; Zhang et al. 2019).

The genus *Terminalia* is well known for its therapeutic values as it has been useful as a cardioprotective, hypolipidemic, antiinflammatory, antioxidant, gastroprotective and wound healing agent. *T. arjuna* is a remarkable plant for treating cardiovascular disorders as it offers multiple modes of cardioprotection. The triterpene glycosides in *T. arjuna* seem to be responsible for its ionotropic effects, whereas the flavonoids and phenolics provide antioxidant effects resulting in cardioprotection (Amalraj and Gopi 2017). Pentacyclic triterpenoids improve epidermal barrier function and induce collagen production and *T. arjuna* bark exerts versatile antiaging properties *in vitro* and *in vivo* (Farwick et al. 2013).

The antidiabetic activity of *T. chebula* has been linked to α-glucosidase inhibition of corilagin and ellagic acid (Li et al. 2014), whereas that of *T. bellirica* is attributed to gallic acid (Latha and Daisy 2011). Gallic acid also showed hypolipidemic activity, which was more pronounced in the *T. arjuna* extract than in the extracts of *T. bellirica* and *T. chebula* (Shaila et al. 1998). A comparative study revealed that *T. arjuna*, *T. chebula* and *T. bellirica* have remarkable antioxidant activities (Chaudhari and Mahajan 2015). An aqueous extract of *T. catappa* nuts also showed antioxidant activity (Krishnaveni 2014). It was found that the antioxidant activities of *T. arjuna*, *T. bellirica* and *T. chebula* were due to their high phenolic contents (Bajpai et al. 2005). Triterpenoids are principally responsible for their cardiovascular properties. Arjunolic acid, chebulic acid, neochebulic acid and corilagin were found to be associated with the hepatoprotective activity (Manna et al. 2007b; Lee et al. 2007; Kinoshita et al. 2007).

Similarly, chebulinic acid was found to be associated with gastroprotective activity (Mishra et al. 2013). Chebulagic acid, anolignan B, punicalagin, ursolic acid and 2α, 3β, 23-trihydroxyurs-12-en-28-oic acid were found to be responsible for the antiinflammatory activity (Nair et al. 2010; Eldeen et al. 2006; Lin et al. 1999; Fan et al. 2004). A recent study revealed that chebulanin from *T. chebula* significantly improved the severity of arthritis and that it is a strong therapeutic alternative for the treatment of rheumatoid arthritis (Zhao et al. 2015).

Casuarinin and punicalagin showed strong antiviral activity (Cheng et al. 2002). Strong antifungal activity was observed with 5,7,2′-tri-O-methyl-flavanone-4′-O-α-L-rhamnopyranosyl-(1→4)-β-D-gluco-pyranoside and 2α, 3β, 19β, 23-tetrahydroxyolean-12-en-28-oicacid-3-O-β-D-galactopyranosyl-(1→3)-β-D-glucopyranoside-28-O-β-D-glucopyranoside isolated from the roots of *T. alata* (Srivastava et al. 2001). Ellagic acid is a potent antioxidant as well as anticarcinogen and has the ability to cause apoptosis in cancer cells (Pfundstein et al. 2010; Saleem et al. 2002). Cytotoxic activity was associated with chebulanic acid, tannic acid, ellagic acid, punicalagin, luteolin, gallic acid and its ethyl ester (Saleem et al. 2002; Reddy et al. 2009; Chen et al. 2000; Pettit et al. 1996). The astringent tannins helped in wound closing and healing processes (Chaudhari and Mengi 2006). The anthraquinone glycosides present in the pericarp of the *T. chebula* fruit are responsible for the laxative effect (Walia and Arora 2013). Arjunolic acid is a novel phytomedicine with multifunctional therapeutic applications (Hemalatha et al. 2010). Among the four phytoconstituents, arjunic acid, arjungenin, arjunetin and arjunglucoside isolated from *T. arjuna*, arjunic acid and its semi-synthetic ester derivatives were found to be significantly active against the human oral (KB), ovarian (PA 1) and liver (HepG-2 and WL-68) cancer cell lines (Saxena et al. 2007). A wide spectrum of biological activities have been reported for betulinic acid, the most important being anticancer (Moghaddam et al. 2012; Hordyjewska et al. 2019). *T. paniculata* exhibited high scavenging efficacy in the case of nitric oxide and hydroxyl radicals (Agrawal et al. 2011).

1.5 PHYTOCHEMICAL ANALYSIS

Several analytical methods, such as high-performance thin-layer chromatography (HPTLC), high-performance liquid chromatography (HPLC) and gas chromatography-mass spectrometry(GC-MS) have been used for the determination of phenolic compounds including phenolic acids, flavonoids, triterpene acids, oleane derivatives and so on (Khatoon et al. 2008; Kumar et al. 2010; Kalola and Rajani 2006; Singh et al. 2002). A precise and rapid high-performance thin-layer chromatography (HPTLC) method was developed for the quantitative determination of four saponins (arjunglucoside-I, arjunetin, arjungenin and arjunolic acid) from fruits of *T. chebula* (Rumalla et al. 2010). From the stem-bark of *T. arjuna*, ellagic acid and gallic acid were quantified by thin-layer chromatography (TLC) and arjunic acid, arjunolic acid, arjungenin, arjunetin and arjunglucoside I were quantified by HPTLC and TLC densitometry (Khatoon et al. 2008; Kumar et al. 2010; Kalola and Rajani 2006; Singh et al. 2002; Uthirapathy and Ahamad 2019). GC-MS analysis identified sixty-four constituents in the ethyl acetate extract of *T. chebula* fruits (Singh and Kumar 2013). Eight hydrolyzable tannins were successfully isolated with high purities (>98%) using preparative HPLC along with UV detection (Mahajan and Pai 2010). Gallic acid, corilagin, chebulagic acid, ellagic acid and chebulinic acid in the extracts of bark and fruits of four *Terminalia* species *T. arjuna*, *T. chebula*, *T. bellirica* and *T. catappa* were quantified by a validated HPLC-PDA method (Dhanani et al. 2015). An RP-HPLC method was developed for determining 14 components, namely gallic acid, chebulic acid, 1,6-di-O-galloyl-D-glucose, punicalagin, 3,4,6-tri-O-galloyl-D-glucose, casuarinin, chebulanin, corilagin, neochebulinic acid, terchebulin, ellagic acid, chebulagic acid, chebulinic acid and 1,2,3,4,6-penta-O-galloyl-D-glucose (Juang et al. 2004). However, these methods suffer from low sensitivity, low resolution and/or long analysis time with the consumption of a large amount of solvents.

A very useful technique for the analysis of plant metabolites is the versatile liquid chromatography-mass spectrometry or LC-MS (Pfundstein et al. 2010). There are a few methods using LC-MS for the analysis of phytocomponents from *Terminalia* species (Avula et al. 2013, 2017; Mopuri et al. 2015; Rane et al. 2016; Venkatalakshmi et al. 2016b; Sobeh et al. 2019). HPLC and LC-IT-MS was used to characterize polyphenols in *T. arjuna* aqueous extracts (Saha et al. 2012). *T. arjuna* bark extracts were analyzed by LC-QTOF-MS and LC-QQQ-MS identifying several polyphenols and quantifying terpenoids arjunic acid, arjungenin and arjunetin (Sekhar et al. 2017; Rane et al. 2016). Fifty components were identified in the leaf extract of *T. bellirica* by liquid chromatography-electrospray ionization mass spectrometry (LC-ESI-MS) in negative ion mode (Sobeh et al. 2019). The ethyl acetate fraction of ethanol extract of *T. bellirica* on LC-MS analysis led to the identification of 14 tannins (Chen et al. 2019). Triphala, a polyherbal formulation in Ayurveda, was analyzed by HPLC and HPLC-ESI-MS (in positive and negative ion mode) to identify the main constituents (Charoenchai et al. 2016). LC-MS analysis of the tannin-rich fraction of hydroalcoholic extract of *T. chebula* showed the presence of gallic acid, methyl gallate, corilagin, chebulagic acid and chebulinic acid (Ekambaram et al. 2018). Positive ion and negative ion LC-MS analysis of *T. chebula* fruit aqueous ethanol extract prepared by ultrasonic-assisted extraction led to the identification of six phenolic compounds, namely shikimic acid, gallic acid, 5-O-galloylshikimic acid, corilagin, 3,4,8,9,10-pentahydroxydibenzo[b,d]pyran-6-one and ellagic acid (Sheng et al. 2018). LC-MS analysis of the methanolic extract of the leaves of *T. coriacea* showing strong antioxidant activity revealed the presence of seven flavonoid glycosides (Khan et al. 2017). LC-MS/MS analysis of the ethanolic extract of *T. paniculata* bark showed the presence of ellagic acid, 2′,4′,5,7, tetramethoxy-8-methylflavanone, 3,3′di-O-methyl ellagic acid, arjunolic acid, galloylarjunolic acid, termilignan and betulinic acid (Mopuri et al. 2015). LC-ESI-MS/MS analysis revealed the presence of isorhamnetin, chlorogenic acid, rottlerin, chrysoeriol, morin, limocitrin, peltatoside, thermoposide, iridin, in the bark, quercetin-3-glucuronide, quercetin-3,4′-O-di-β-glucopyranoside, geniposide, rutin, hesperitin, flavanomarein, kaempferol-7-neohesperioside, baccatin, isorhamnetin, peonidin, iristectorin-A, scoparin, tricin, cursiliol, isorhamnetin-3-glucoside-4′-glucoside in fruit and mucic acid, quercetin-7-O-rhamnoside, limocitrin, rottlerin, sophoricoside, hesperitin, quercetin-3-D-xyloside, isorhamnetin, iristectorin-A, kaempferol-3-O-alpha-L-arabinoside and myricetin-3-galactoside in wood of *T. catappa* (Venkatalakshmi et al. 2016b). The components absorbed in the plasma and brain tissue of rats after intragastric administration of a *T. chebula* extract were analyzed by UPLC-QTOF-MS identifying nine compounds in the plasma and five in the brain (Zhang et al. 2018).

Identification of Bioactive Phytoconstituents of *Terminalia* Species by HPLC-ESI-QTOF-MS/MS

2

2.1 INTRODUCTION

There is no comprehensive study on the identification and characterization of the phytoconstituents in different plant parts of *Terminalia* species. LC-QTOF-MS gives separation of the components, accurate mass and MS/MS data enabling identification and characterization of the components. We, therefore, used LC-QTOF-MS to profile, identify and characterize the phytocomponents of six species of *Terminalia* (*T. arjuna*, *T. bellirica*, *T. catappa*, *T. chebula*, *T. elliptica* and *T. paniculata*) available in India (Singh et al. 2016a, 2018b). A large number of compounds (179) including acids, flavonoids, gallic and ellagic acid derivatives and proanthocyanidins were identified and characterized by comparison with reference compounds, interpretation of MS/MS fragmentation patterns or by comparison with literature data.

2.2 PLANT MATERIALS AND CHEMICALS

Bark, fruit, leaf, stem, and root of the six *Terminalia* species were collected from Jawaharlal Nehru Tropical Botanic Garden and Research Institute campus (JNTBGRI; N: 8°45', E:77°10', Altitude: 70–160 m), Kerala, South India in the month of February 2014.

Voucher herbarium specimen numbers of *T. arjuna* (Roxb. ex. DC) Wight & Arn., *T. bellirica* (Gaertner) Roxb., *T. catappa* L., *T. chebula* Retz., *T. elliptica* Willd. and *T. paniculata* Roth were 66460, 66461, 66464, 66462, 66465 and 66463, respectively, and deposited in the Herbarium of JNTBGRI (TBGT).

AR grade ethanol (Merck, Darmstadt, Germany) was used to prepare the ethanolic extracts. LC-MS grade methanol, acetonitrile and formic acid (Sigma-Aldrich, St. Louis, MO, USA), were used in the mobile phase and sample preparation throughout the LC-MS studies. Ultra-pure water, obtained from the Direct-Q water purification system (Millipore, Billerica, MA, USA), was used throughout the analysis.

The reference standards of (+)-catechin, (+)-epicatechin, mangiferin, kaempferol-3-*O*-rutinoside, quercetin-3,4'-di-*O*-glucoside, rutin, ellagic acid, gallic acid, naringin, genistein, amentoflavone, chrysin and protocatechuic acid were purchased from Extrasynthese (Genay, France). The reference standards of orientin, isoorientin, vitexin, isovitexin, kaempferol, quercetin, luteolin, apigenin, eriodictyol, scutellarein, quinic acid, ferulic acid, sinapic acid, vanillic acid, caffeic acid, chlorogenic acid, betulinic acid, oleanolic acid, ursolic acid, arjunolic acid, arjungenin, arjunetin and chebulinic acid were purchased from Sigma Aldrich Ltd. (St. Louis, MO, USA). The reference standard chebulagic acid was purchased from Natural Remedies Pvt. Ltd. (Bangalore, India). The purity of all the reference standards was ≥95%.

2.3 EXTRACTION AND SAMPLE PREPARATION

Each plant part (leaf, stem, root, fruit and bark) of the six *Terminalia* species was washed, dried and powdered and 50 g was extracted with 500 mL of ethanol (100%) in an ultrasonic water bath for 15 min and then kept at room temperature. The extracts were filtered after 24 h, through filter paper (Whatman No. 1) and residues were re-extracted three times with fresh solvent following the same procedure and the filtrates combined. Each sample filtrate was evaporated to dryness under reduced pressure at 20–50 kPa and temperature at 40°C using a Buchi rotary evaporator (Flawil, Switzerland). Using methanol, 1 mg/mL stock solution of the extract of each sample was prepared separately and filtered through a 0.22 μm PVDF membrane (Merck Millipore, Darmstadt, Germany) and stored at −20°C. The stock solution was suitably diluted with methanol to prepare the final working concentrations. A stock solution of each selected reference standard was prepared in methanol (1,000 μg/mL) which was suitably diluted with methanol to prepare the final working concentrations.

2.4 HPLC-ESI-QTOF-MS/MS CONDITIONS

An Agilent 6520 QTOF-MS/MS system hyphenated with an Agilent 1200 HPLC system (Agilent Technologies; Santa Clara, CA, USA) via an ESI interface was used for the screening of phytoconstituents from ethanolic extracts of *Terminalia* species. The separation was achieved on a Thermo Betasil C8 column (250 mm×4.5 mm, 5 μ) at 25°C with 0.1% formic acid aqueous solution (A) and acetonitrile (B) in a gradient elution (B) at 0.5 mL/min using the program 10%–20% (B) from 1.0 to 12 min, 20%–25% (B) from 12 to 20 min, 25%–30% (B) from 20 to 25 min, 30%–35% from 25 to 28 min, 35%–40% (B) from 28 to 32 min, 40%–70% (B) from 32 to 45 min and return to initial conditions over 10 min with UV detection at 200–400 nm. The mass spectrometer was operated in negative ESI mode at a resolving power >15,000 (FWHM) in a mass range of m/z 50–2,000. Nitrogen was used as drying and collision gas. The heated capillary temperature was set to 350°C and the nebulizer pressure was set to 35 psi. The drying gas flow rate was 10 L/min. VCap, fragmentor, skimmer and octopole RF peak voltages were set to 3,500, 150, 65 and 750 V, respectively in negative polarity. The MS/MS spectra were acquired by auto-fragmentation where the three most intense mass peaks where fragmented. Collision energy values for MS/MS experiments were fixed at 20–40 eV for each selected mass. The chromatographic and mass spectrometric analyses, including the prediction of chemical formula and exact mass calculation, were performed by using the Mass Hunter software version B.04.00 build 4.0.479.0 (Agilent Technology).

2.5 QUALITATIVE ANALYSIS OF THE PHYTOCONSTITUENTS OF *TERMINALIA* SPECIES

2.5.1 LC-MS Analysis

The working solution of *T. chebula* fruit extract was used to optimize the sensitivity and resolution and was initially analyzed in both positive and negative ionization modes. Negative ion mode was found more sensitive toward the phytoconstituents in the sample. Therefore, further analysis was done in negative ionization mode resulting in deprotonated molecular ion peaks. The LC-MS analysis was carried out for the bark, fruit, leaf, stem and root extracts of the six *Terminalia* species (30 samples) under the same optimized conditions. The resulting base peak chromatograms (BPCs) are shown in Figure 2.1. A total of 179 compounds including 25 acids, 35 flavonoids, 27 gallic acid (GA) derivatives, 49 ellagic acid (EA) derivatives and 43 proanthocyanidins (PAs) extracted from their BPCs using a mass tolerance window of ±5 ppm and their respective peak retention times (RT) are reported in Table 2.1. The mass spectra derived from these EICs showed abundant [M-H]⁻ ions with a mass error <2.46 ppm. For further characterization, these ions were subjected to MS/MS analysis at varied collision energies of 20–40 eV, under collision-induced dissociation (CID) conditions.

2.5.2 Identification of Acids

A total of 25 peaks were detected as organic acids in different plant parts of *Terminalia* species (Table 2.1). Peaks **1, 25, 64, 72, 112, 128, 129, 138, 150, 158, 162, 167, 170, 175–177** and **179** were positively identified by comparing their RT, MS and MS/MS pattern with those of authentic reference standards. Peaks **2, 3, 6, 13, 42, 74, 148** and **160** were identified and characterized on the basis of their exact masses and MS/MS fragmentation patterns as shikimic acid, salicylic acid, mallic acid, melilotic acid, methyl protocatechuate, 3-*O*-*p*-coumaroylquinic acid, *p*-coumaric acid and ethyl protocatechuate, respectively.

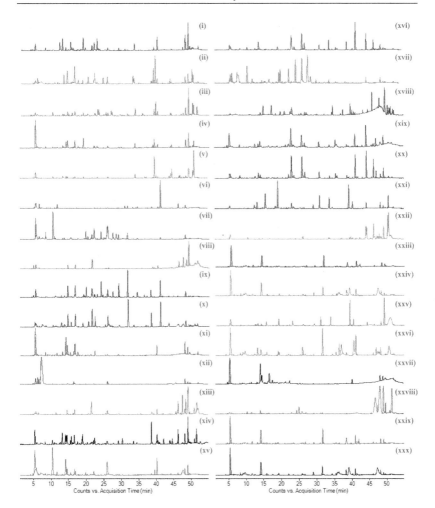

FIGURE 2.1 Base peak chromatograms of (i) *T. arjuna (TAJ)* bark; (ii) *T. arjuna* fruit; (iii) *T. arjuna* leaf; (iv) *T. arjuna* stem; (v) *T. arjuna* root; (vi) *T. bellirica (TBL)* bark; (vii) *T. bellirica* fruit; (viii) *T. bellirica* leaf; (ix) *T. bellirica* stem; (x) *T. bellirica* root; (xi) *T. catappa (TCT)* bark; (xii) *T. catappa* fruit; (xiii) *T. catappa* leaf; (xiv) *T. catappa* stem; (xv) *T. catappa* root; (xvi) *T. chebula (TCB)* bark; (xvii) *T. chebula* fruit; (xviii) *T. chebula* leaf; (xix) *T. chebula* stem; (xx) *T. chebula* root; (xxi) *T. elliptica (TEP)* bark; (xxii) *T. elliptica* fruit; (xxiii) *T. elliptica* leaf; (xxiv) *T. elliptica* stem; (xxv) *T. elliptica* root; (xxvi) *T. paniculata (TPN)* bark; (xxvii) *T. paniculata* fruit; (xxviii) *T. paniculata* leaf (xxix) *T. paniculata* stem; (xxx) *T. paniculata* root.

TABLE 2.1 HPLC-ESI-QTOF-MS and CID-MS/MS data of phytoconstituents from *Terminalia* species

PEAK NO.	RT (IN MIN)	MOLECULAR FORMULA	[M-H]− (CALC)	[M-H]− (OBS)	ERROR (ΔPPM)	MS/MS IN NEGATIVE ION MODE	IDENTIFICATION	*DETECTION IN SAMPLES	**REF.
Identification of acids									
1	5.5	$C_7H_{12}O_6$	191.0561	191.0561	0.00	127.04, 109.03, 93.04, 85.03 (100), 81.04, 59.02, 43.02	Quinic acid[a]	1–30	1
2	5.8	$C_7H_{10}O_5$	173.0455	173.0453	−1.16	155.04, 137.03, 111.05, 99.05, 93.04 (100), 83.05, 73.0307, 65.04	Shikimic acid	1–20, 22–25, 27–30	2
3	6.1	$C_7H_6O_3$	137.0244	137.0243	−0.73	93.03 (100), 65.04	Salicylic acid	2, 3, 24, 28	3
6	6.2	$C_4H_6O_5$	133.0142	133.0142	0.00	115.00 (100)	Malic acid	1–30	4, 5
13	8.3	$C_9H_{10}O_3$	165.0557	165.0556	−0.61	147.04, 121.03, 119.05 (100), 93.03, 77.04	Mellilotic acid	3–5, 8–12, 14, 17–27	–
25	9.9	$C_7H_6O_5$	169.0142	169.0141	−0.59	125.0250 (100), 107.0149, 97.0307, 79.0202, 69.0359, 51.0255	Gallic acid[a]	1–30	5, 6
42	13.4	$C_8H_8O_4$	167.035	167.0348	−1.20	152.01, 108.02(100)	Methyl protocatechuate	1, 18, 29	–
64	16.6	$C_7H_6O_4$	153.0193	153.0192	−0.65	109.03(100), 91.02, 81.04, 53.04, 44.99	Protocatechuic acid[a]	1–30	5, 7
72	18	$C_{16}H_{18}O_9$	353.0878	353.0871	−1.98	191.06 (100), 161.02	Chlorogenic acid[a]	3, 4, 13, 17–21, 29	8

(Continued)

TABLE 2.1 (Continued) HPLC-ESI-QTOF-MS and CID-MS/MS data of phytoconstituents from *Terminalia* species

PEAK NO.	RT (IN MIN)	MOLECULAR FORMULA	[M-H]⁻ (CALC)	[M-H]⁻ (OBS)	ERROR (ΔPPM)	MS/MS IN NEGATIVE ION MODE	IDENTIFICATION	*DETECTION IN SAMPLES	**REF.
74	18.1	$C_{16}H_{18}O_8$	337.0929	337.0929	0.00	191.05, 173.04, 163.04(100), 155.03, 119.05, 111.04, 93.03	3-O-p-coumaroyl quinic acid	1–10, 12–18, 29, 30	5
112	23.1	$C_9H_8O_4$	179.035	179.0347	−1.68	135.0448 (100), 117.0341 (4)	Caffeic acid[a]	3, 4, 8, 9, 13, 18–20, 23, 26–30	9
128	24.9	$C_{41}H_{30}O_{27}$	953.0902	953.09	−0.21	935.08, 853.07, 783.07, 633.07, 481.06, 463.05, 419.06, 337.02, 319.01, 300.99 (100), 275.02, 205.05	Chebulagic acid[a]	1–4, 6–24, 26–30	–
129	25.3	$C_{14}H_6O_8$	300.999	300.999	0.00	283.99, 245.01, 229.01, 216.01, 200.01, 185.02, 145.03 (100), 129.03	Ellagic acid[a]	1–30	6
138	27.3	$C_{41}H_{32}O_{27}$	955.1058	955.105	−0.84	937.10, 911.11, 855.11, 803.10, 785.09, 767.07, 741.11, 685.07, 633.07, 617.08, 465.07, 447.06, 337.02 (100), 319.01, 293.03, 275.02, 205.05, 169.02	Chebulinic acid[a]	1–30	–
148	29.4	$C_9H_8O_3$	163.0401	163.0402	0.61	119.05 (100), 117.03, 93.03	P-coumaric acid	2–5, 16, 18, 20, 28	5, 7

(Continued)

TABLE 2.1 (Continued)   HPLC-ESI-QTOF-MS and CID-MS/MS data of phytoconstituents from *Terminalia* species

PEAK NO.	RT (IN MIN)	MOLECULAR FORMULA	[M-H]⁻ (CALC)	[M-H]⁻ (OBS)	ERROR (ΔPPM)	MS/MS IN NEGATIVE ION MODE	IDENTIFICATION	*DETECTION IN SAMPLES	**REF.
150	29.8	$C_{11}H_{12}O_5$	223.0612	223.0609	−1.34	208.04 (100), 193.02, 179.07, 164.05, 149.03, 135.05, 121.03	Sinapic acid[a]	1–5, 7, 9–17, 19–30	10
158	35.1	$C_8H_8O_4$	167.035	167.0349	−0.60	151.00, 123.04, 107.01, 95.05, 83.01, 81.03(100), 63.03, 57.04	Vanillic acid[a]	1–11, 14–16, 21–30	5
160	37.7	$C_9H_{10}O_4$	181.0506	181.0503	−1.66	153.02, 108.02 (100), 92.92	Ethyl protocatechuate	1, 5–30	—
162	39.5	$C_{36}H_{58}O_{10}$	649.3957	649.3945	−1.85	487.35 (100)	Arjunetin[a]	1–30	—
		$C_{37}H_{59}O_{12}$	695.4012 [M+COOH]⁻			649.40, 487.35 (100), 469.33, 207.05	Arjunetin adduct	1–30	—
167	41.6	$C_{10}H_{10}O_4$	193.0506	193.0505	−0.52	178.03, 149.03, 134.04 (100)	Ferulic acid[a]	1–4, 6–10, 13–30	—
170	43.8	$C_{30}H_{48}O_6$	503.3378	503.337	−1.59	485.33, 457.33, 453.30, 441.34, 435.29, 425.31, 421.31, 409.31 (100), 407.30, 403.30, 393.31, 391.30, 379.30, 363.27	Arjungenin[a]	1–30	—
175	48.1	$C_{30}H_{48}O_3$	455.3531	455.353	−0.22	407.34	Betulinic acid[a]	1–6, 8–30	—
176	48.5	$C_{30}H_{48}O_3$	455.3531	455.3535	0.88	407.33, 391.29, 384.93, 373.26	Oleanolic acid[a]	1–6, 8–16, 18–20, 22–30	11

(Continued)

TABLE 2.1 (Continued) HPLC-ESI-QTOF-MS and CID-MS/MS data of phytoconstituents from *Terminalia* species

PEAK NO.	RT (IN MIN)	MOLECULAR FORMULA	[M-H]⁻ (CALC)	[M-H]⁻ (OBS)	ERROR (ΔPPM)	MS/MS IN NEGATIVE ION MODE	IDENTIFICATION	*DETECTION IN SAMPLES	**REF.
177	49	$C_{30}H_{48}O_5$	487.3429	487.3428	−0.21	421.31, 409.31, 407.29, 403.30, 393.32, 391.29, 377.28, 363.27	Arjunolic acid[a]	1–6, 8–25, 27–30	–
179	49.9	$C_{30}H_{48}O_3$	455.3531	455.3528	−0.66	407.34, 325.18, 183.01	Ursolic acid[a]	1–5, 8–15, 18–29	12
Identification of flavonoids									
39	13	$C_{15}H_{14}O_7$	305.0667	305.0667	0.00	261.07, 243.06, 237.08, 219.06, 203.03, 191.07, 179.03, 167.03, 165.02, 139.04, 137.02, 125.02(100), 111.04, 109.03	(+)-Gallocatechin	1–30	6
54	15.7	$C_{15}H_{14}O_7$	305.0667	305.0667	0.00	261.07, 243.06, 237.08, 219.06, 203.03, 191.07, 179.03, 167.03, 165.02, 139.04, 137.02(30), 125.02(100), 111.04, 109.03	(−)-Epigallocatechin	1–30	6
80	19.2	$C_{15}H_{14}O_6$	289.0718	289.0718	0.00	271.06, 245.08, 221.08, 203.07, 187.04, 161.06, 151.04, 137.02, 125.02, 123.04, 121.03, 109.03 (100), 97.03, 81.03	(+)-Catechin[a]	1–30	6

(Continued)

TABLE 2.1 (Continued)   HPLC-ESI-QTOF-MS and CID-MS/MS data of phytoconstituents from *Terminalia* species

PEAK NO.	RT (IN MIN)	MOLECULAR FORMULA	[M-H]⁻ (CALC)	[M-H]⁻ (OBS)	ERROR (ΔPPM)	MS/MS IN NEGATIVE ION MODE	IDENTIFICATION	*DETECTION IN SAMPLES	**REF.
93	20.9	$C_{21}H_{20}O_{11}$	447.0933	447.0935	0.45	413.12, 357.06, 339.05, 327.05 (100), 299.04	Orientin[a]	2, 3, 6, 9, 12, 13, 16–20, 22–26, 27, 29	13
95	21.2	$C_{21}H_{20}O_{11}$	447.0933	447.0934	0.22	301.04(100), 300.03, 271.02, 255.03, 179.00, 151.00	Quercetin-3-O-rhamnoside	3, 16–21, 28	1
96	21.2	$C_{28}H_{24}O_{11}$	631.0941	631.094	−0.16	479.08 (100), 316.02, 299.02	Myricetin galloyl hexoside	2, 7, 18, 19, 27	1
98	21.3	$C_{15}H_{14}O_{6}$	289.0718	289.0717	−0.35	271.06, 245.08, 221.08, 203.07, 187.04, 161.06, 151.04, 137.02, 125.02, 123.04, 121.03, 109.03(100), 97.03, 81.03	(–)-Epicatechin[a]	1–30	6, 7
99	21.3	$C_{27}H_{30}O_{17}$	625.141	625.141	0.00	463.09, 301.03 (100)	Quercetin-3, 4'-di-O- glucoside[a]	1–4, 6, 7, 9, 11–30	—
100	21.4	$C_{21}H_{20}O_{11}$	447.0933	447.0931	−0.45	429.07, 411.04, 357.06 (100), 353.05, 339.05, 327.04, 300.03	Isoorientin[a]	1–30	—
109	22.6	$C_{21}H_{20}O_{13}$	479.0831	479.0827	−0.83	316.02(100), 287.02, 271.03, 178.99	Myricetin galactoside	1–6, 11, 13–15, 18, 25, 26, 28–30	1

(Continued)

TABLE 2.1 (*Continued*) HPLC-ESI-QTOF-MS and CID-MS/MS data of phytoconstituents from *Terminalia* species

PEAK NO.	RT (IN MIN)	MOLECULAR FORMULA	[M−H]⁻ (CALC)	[M−H]⁻ (OBS)	ERROR (ΔPPM)	MS/MS IN NEGATIVE ION MODE	IDENTIFICATION	*DETECTION IN SAMPLES	**REF.
111	23	$C_{21}H_{20}O_{13}$	479.0831	479.0831	0.00	316.02(100), 287.02, 271.03, 178.99	Myricetin glucoside	1–5, 11, 13–15, 18, 25, 26, 28–30	14
120	24.1	$C_{21}H_{18}O_{13}$	477.0675	477.0674	−0.21	462.05, 315.01 (100), 299.99	3-O-methyl quercetin-7-O-glucopyranoside	1–4, 6–11, 13, 14, 16, 19, 21, 25, 26, 28–30	15
121	24.2	$C_{27}H_{30}O_{16}$	609.1461	609.1462	0.16	300.03 (100), 301.03, 178.99	Rutin[a]	1–4, 8, 9, 11–14, 16–20, 22–24, 26–30	1, 16
122	24.2	$C_{28}H_{24}O_{16}$	615.0992	615.0994	0.33	463.09 (100), 313.06, 301.04, 300.03, 271.05	Myricetin galloyl rhamnopyranoside	2, 13, 14, 28	1
125	24.8	$C_{21}H_{20}O_{10}$	431.0984	431.0992	1.86	413.09, 353.07, 341.07, 323.06, 311.06 (100), 295.06, 283.06, 269.05	Vitexin[a]	1–6, 8–10, 12–29	7
126	24.8	$C_{28}H_{24}O_{16}$	615.0992	615.0991	−0.16	463.09 (100), 313.06, 301.04, 300.03, 271.05	Quercetin-3-O-galloyl-hexoside	2, 10, 13, 14, 18, 28	17
131	25.9	$C_{21}H_{20}O_{10}$	431.0984	431.098	−0.93	413.08, 396.08, 371.07, 353.07, 341.06, 311.06 (100), 283.06, 243.07	Isovitexin[a]	2, 3, 13, 17–20, 22–24, 27, 29	–
132	26.2	$C_{21}H_{20}O_{12}$	463.0882	463.0881	−0.22	316.02(100), 300.02, 271.02, 178.99	Myricetin-3-O-rhamnoside	1–5, 7, 9–30	1
133	26.3	$C_{21}H_{20}O_{12}$	463.0882	463.0881	−0.22	301.04, 300.03 (100)	Isoquercitrin	2–4, 7, 9, 11–30	7, 16

(*Continued*)

TABLE 2.1 (Continued)   HPLC-ESI-QTOF-MS and CID-MS/MS data of phytoconstituents from *Terminalia* species

PEAK NO.	RT (IN MIN)	MOLECULAR FORMULA	[M-H]⁻ (CALC)	[M-H]⁻ (OBS)	ERROR (ΔPPM)	MS/MS IN NEGATIVE ION MODE	IDENTIFICATION	*DETECTION IN SAMPLES	**REF.
134	26.3	$C_{27}H_{30}O_{15}$	593.1512	593.1518	1.01	301.034 (100), 289.07, 285.04	Kaempferol-3-O-rutinoside[a]	1–10, 12–14, 16–20, 23–25, 27, 28, 30	4
136	26.6	$C_{21}H_{20}O_{12}$	463.0882	463.0881	−0.22	301.04 (67), 300.03 (100)	Hyperoside	1–30	7
145	28.8	$C_{28}H_{24}O_{15}$	599.1042	599.1041	−0.17	447.09, 313.06, 285.04, 241.04, 169.01	Kaempferol-3-O-galloyl-hexoside	2, 18, 28	1
149	29.5	$C_{21}H_{20}O_{11}$	447.0933	447.0935	0.45	301.04, 300.03 (100), 271.02, 255.03, 178.99, 151.00	Quercetin deoxyhexoside	3, 4, 22, 28	18
154	30.9	$C_{27}H_{32}O_{14}$	579.1719	579.171	−1.55	459.11, 313.07, 271.06 (100), 150.99	Naringin[a]	2–4, 6–8, 17, 20, 23–25, 27, 30	–
157	35.1	$C_{15}H_{10}O_8$	317.0303	317.0301	−0.63	193.014, 178.99, 151.00(100), 137.02, 125.02, 109.03, 107.01	Myricetin	1–30	19
159	36.5	$C_{15}H_{10}O_6$	285.0405	285.0398	−2.46	137.03, 117.04(100)	Scutellarein[a]	1, 5–10, 12, 13, 16–28, 30	
163	40	$C_{15}H_{10}O_6$	285.0405	285.0405	0.00	267.03, 217.05, 199.04, 175.04, 151.00, 133.03, 121.03, 107.02	Luteolin[a]	3, 8, 9, 13, 18, 20, 23, 24, 27–29,	5, 20
164	40.1	$C_{15}H_{10}O_7$	301.0354	301.0352	−0.66	273.04, 178.99, 151.00 (100), 121.03, 107.01, 83.01, 65.00	Quercetine[a]	1, 3, 5–10, 12–14, 17, 19, 20, 22, 28	1, 20

(Continued)

TABLE 2.1 (Continued) HPLC-ESI-QTOF-MS and CID-MS/MS data of phytoconstituents from *Terminalia* species

PEAK NO.	RT (IN MIN)	MOLECULAR FORMULA	$[M-H]^-$ (CALC)	$[M-H]^-$ (OBS)	ERROR (ΔPPM)	MS/MS IN NEGATIVE ION MODE	IDENTIFICATION	*DETECTION IN SAMPLES	**REF.
165	40.4	$C_{15}H_{12}O_6$	287.0561	287.0561	0.00	151.00 (100), 135.04, 125.02, 107.01	Eriodictyol[a]	1–10, 12–30	20
169	43.4	$C_{15}H_{10}O_5$	269.0455	269.0452	−1.12	224.04, 196.06, 183.05, 180.09, 159.07, 157.07, 135.05, 133.06 (100), 107.05, 91.05, 63.07	Genistein[a]	1–30	–
171	44.1	$C_{15}H_{10}O_6$	285.0405	285.0401	−1.40	151.01, 133.03 (100)	Kaempferol[a]	1–9, 11–20, 22–30	–
172	44.3	$C_{15}H_{10}O_5$	269.0455	269.0451	−1.49	151.00, 149.02, 117.03 (100), 107.01	Apigenin[a]	1–30	–
173	44.6	$C_{30}H_{18}O_{10}$	537.0827	537.0821	−1.12	417.05, 375.05 (100)	Amentoflavone[a]	8, 9, 17, 18, 20, 23	–
174	46.6	$C_{17}H_{14}O_6$	313.0718	313.0718	0.00	298.04, 297.04, 283.02(100), 269.04, 255.03	Dihydroxy dimethoxyflavone	2, 15	21
178	49.3	$C_{15}H_{10}O_4$	253.0506	253.0506	0.00	167.05, 145.03, 143.03 (100), 119.05, 115.06, 107.01, 101.04, 89.00	Chrysin[a]	1–5, 8–14, 16–25, 27, 28	–
Identification of gallic acid derivatives									
5	6.2	$C_{13}H_{16}O_{10}$	331.0671	331.0671	0.00	271.05, 211.03, 169.02 (100), 151.002, 125.03, 113.03, 71.02	Galloylglucose Iso I	1–30	6, 22

(Continued)

TABLE 2.1 (Continued)   HPLC-ESI-QTOF-MS and CID-MS/MS data of phytoconstituents from *Terminalia* species

PEAK NO.	RT (IN MIN)	MOLECULAR FORMULA	$[M-H]^-$ (CALC)	$[M-H]^-$ (OBS)	ERROR (ΔPPM)	MS/MS IN NEGATIVE ION MODE	IDENTIFICATION	*DETECTION IN SAMPLES	**REF.
11	8.1	$C_{13}H_{16}O_{10}$	331.0671	331.0672	0.30	271.05, 211.03, 169.01(100), 151.00, 125.03, 113.03, 71.02	Galloylglucose Iso II	1–30	6, 22
18	9	$C_{20}H_{20}O_{14}$	483.078	483.0782	0.41	331.07, 313.06, 271.05, 211.02, 169.01 (100), 125.03	Digalloylglucose Iso I	2–5, 7–13, 16, 19, 21–26, 30	22, 23
22	9.7	$C_{20}H_{20}O_{14}$	483.078	483.0783	0.62	331.07, 313.06, 271.05 (100), 211.03, 169.02, 125.03	Digalloylglucose Iso II	2–5, 7–13, 16, 19, 21–26, 30	22, 23
31	11.8	$C_{14}H_{14}O_9$	325.0565	325.0565	0.00	169.02 (100), 137.03, 125.03, 111.05, 93.04	Galloylshikimic acid Iso I	2–4, 8, 12–19, 25, 28	–
33	12.1	$C_{14}H_{14}O_9$	325.0565	325.0564	−0.31	169.02 (100), 137.03, 125.03, 111.05, 93.04	Galloylshikimic acid Iso II	2–4, 8, 12–19, 25, 28	–
65	17.1	$C_{27}H_{24}O_{18}$	635.089	635.0889	−0.16	483.07, 465.07, 423.05, 300.99, 271.04, 169.01 (100), 125.02	Trigalloylglucose Iso I	2–4, 7–10, 12–20, 22–27	22
68	17.6	$C_{20}H_{22}O_{12}$	453.1038	453.1039	0.22	313.05, 169.01(100), 151.00, 139.04, 125.02, 123.01	3-Methoxy-4-hydroxy phenol-O-(6′-O-galloyl)-glucoside	1–5, 7–18, 20, 22–30	–
79	18.6	$C_{27}H_{24}O_{18}$	635.089	635.089	0.00	483.07, 465.07, 423.05, 300.99, 271.04, 169.01(100), 125.02	Trigalloylglucose Iso II	2–4, 7–10, 12–20, 22–27	22

(Continued)

TABLE 2.1 (Continued) HPLC-ESI-QTOF-MS and CID-MS/MS data of phytoconstituents from *Terminalia* species

PEAK NO.	RT (IN MIN)	MOLECULAR FORMULA	[M-H]⁻ (CALC)	[M-H]⁻ (OBS)	ERROR (ΔPPM)	MS/MS IN NEGATIVE ION MODE	IDENTIFICATION	*DETECTION IN SAMPLES	**REF.
84	19.5	$C_{21}H_{18}O_{13}$	477.0675	477.0675	0.00	325.06, 169.01(100) 125.02, 93.04	Digalloyl-shikimic acid Iso I	2, 3, 8, 10, 12, 17–20, 23, 27, 30	–
89	20.6	$C_{21}H_{18}O_{13}$	477.0675	477.0675	0.00	325.06, 169.01(100) 125.02, 93.04	Digalloyl-shikimic acid IsoII	1–18, 20–30	–
91	20.8	$C_{8}H_{8}O_{5}$	183.0299	183.0299	0.00	168.01, 124.02(100), 78.01	Methyl gallate	1–30	22
92	20.9	$C_{22}H_{18}O_{11}$	457.0776	457.0774	-0.44	305.07, 169.01 (100), 125.02	(+)-Gallocatechin-3-O-gallate	2, 4, 5, 11, 13–15, 21, 25	17
105	22	$C_{22}H_{18}O_{11}$	457.0776	457.0777	0.22	305.07, 169.01(100), 125.02	(−)-Epigallocatechin-3-O-gallate	1–6, 9, 11–16, 21, 25–30	17
106	22.2	$C_{34}H_{28}O_{22}$	787.0999	787.1001	0.25	635.09, 617.08, 483.08, 465.06, 447.06, 313.05, 300.99, 271.05, 169.02 (100), 125.03	Tetragalloylglucose Iso I	2, 7, 11, 14, 15, 19, 20, 22–27, 29	22
119	23.6	$C_{34}H_{28}O_{22}$	787.0999	787.0999	0.00	635.09, 617.08 (100), 483.08, 465.07, 447.06, 313.05, 300.99, 271.04, 169.02, 125.03	Tetragalloylglucose Iso II	2, 7, 11, 14, 15, 19, 20, 22–27, 29	22
123	24.6	$C_{34}H_{28}O_{22}$	787.0999	787.1001	0.25	635.09, 617.08(100), 483.08, 465.06, 447.06, 313.05, 300.99, 271.05, 169.02, 125.03	Tetragalloylglucose Iso III	2–7, 11, 14, 15, 19, 20, 22–27, 29	22

(Continued)

TABLE 2.1 (Continued) HPLC-ESI-QTOF-MS and CID-MS/MS data of phytoconstituents from *Terminalia* species

PEAK NO.	RT (IN MIN)	MOLECULAR FORMULA	[M-H]⁻ (CALC)	[M-H]⁻ (OBS)	ERROR (ΔPPM)	MS/MS IN NEGATIVE ION MODE	IDENTIFICATION	*DETECTION IN SAMPLES	**REF.
130	25.7	$C_{28}H_{22}O_{17}$	629.0784	629.0786	0.32	477.07, 325.06, 289.03, 169.01 (100), 137.02, 125.03, 123.01	Trigalloylshikimic acid	2-4, 7-11, 13, 15-20, 22, 27, 29	22
135	26.4	$C_{34}H_{28}O_{22}$	787.0999	787.1002	0.38	635.09, 617.08 (100), 483.08, 465.06, 447.06, 313.05, 300.99, 271.05, 169.02 (100), 125.03	Tetragalloylglucose Iso IV	2, 7, 11, 14, 15, 19, 20, 22-27, 29	22
137	27.3	$C_{34}H_{28}O_{22}$	787.0999	787.0997	-0.25	635.09, 617.08 (100), 483.08, 465.06, 447.06, 313.05, 300.99, 271.05, 169.02, 125.03	Tetragalloylglucose Iso V	2, 7, 11, 14, 15, 19, 20, 22-27, 29	22
139	27.5	$C_{22}H_{14}O_{12}$	469.0412	469.0409	-0.64	317.04, 169.01 (100), 125.02	Myricetin-3-O-gallate	1-24, 26-30	–
141	28	$C_{41}H_{32}O_{26}$	939.1109	939.111	0.11	787.11, 769.09, 617.08, 465, 447.07, 271.04, 169.02 (100), 125.02	Pentagalloylglucose Iso I	1-30	22
142	28.3	$C_{22}H_{18}O_{10}$	441.0827	441.0826	-0.23	289.07, 245.08, 169.01(100), 137.02, 125.02, 109.03	(+)-Catechin-3-O-gallate	1-30	17
143	28.4	$C_{28}H_{24}O_{14}$	583.1093	583.1091	-0.34	431.09, 341.06, 311.05 (100), 271.04, 269.05, 211.02, 169.01, 125.02	Vitexin-2″-O-gallate	2, 3, 7, 9, 11, 16-21, 26	–

(Continued)

TABLE 2.1 (Continued)   HPLC-ESI-QTOF-MS and CID-MS/MS data of phytoconstituents from *Terminalia* species

PEAK NO.	RT (IN MIN)	MOLECULAR FORMULA	[M-H]⁻ (CALC)	[M-H]⁻ (OBS)	ERROR (ΔPPM)	MS/MS IN NEGATIVE ION MODE	IDENTIFICATION	*DETECTION IN SAMPLES	**REF.
144	28.7	$C_9H_{10}O_5$	197.0455	197.0455	0.00	169.02, 124.02 (100), 78.01	Ethyl gallate	1–30	–
147	29.3	$C_{22}H_{18}O_{10}$	441.0827	441.0827	0.00	289.07, 245.08, 169.01 (100), 137.02, 125.02, 109.03	(−)-Epicatechin-3-O-gallate	2–10, 11, 15–25, 29	17
151	30.3	$C_{41}H_{32}O_{26}$	939.1109	939.1108	−0.11	787.11, 769.09, 617.08, 465, 447.07, 277.04, 169.02 (100)	Pentagalloylglucose Iso II	2, 6, 10, 18, 22, 27	22
Identification of ellagic acid derivatives									
7	6.3	$C_{20}H_{18}O_{14}$	481.0624	481.0625	0.21	421.04, 300.99 (100), 275.02,	HHDP-glucose	1–5, 7–30	23
8	6.3	$C_{27}H_{20}O_{18}$	631.0577	631.0576	−0.16	613.04, 587.06, 569.06, 551.04, 479.03, 467.02, 461.01, 449.01, 441.04, 438.02, 425.01, 423.04, 327.01, 315.01, 300.99, 297.00	Castalin/Vescalin	1, 4, 5, 11, 14–16, 19, 21–30	24
12	8.3	$C_{34}H_{22}O_{22}$	781.053	781.0529	−0.13	721.03, 600.09 (100), 575.01, 392.99, 298.98, 273.00, 270.99	Punicalin β	1–20, 22–25, 27–30	22

(Continued)

TABLE 2.1 (Continued) HPLC-ESI-QTOF-MS and CID-MS/MS data of phytoconstituents from *Terminalia* species

PEAK NO.	RT (IN MIN)	MOLECULAR FORMULA	[M-H]⁻ (CALC)	[M-H]⁻ (OBS)	ERROR (ΔPPM)	MS/MS IN NEGATIVE ION MODE	IDENTIFICATION	*DETECTION IN SAMPLES	**REF.
14	8.4	$C_{27}H_{22}O_{19}$	649.0677	649.0687	1.54	605.08, 481.06, 479.04, 451.05, 425.01, 300.99 (100), 298.99, 275.02, 273.00	Galloyl-HHDP-gluconate Iso I	2, 18, 21	6
15	8.6	$C_{34}H_{24}O_{22}$	783.0686	783.069	0.51	481.06, 300.99 (100), 275.02,	Pedunculagin Iso I (di-HHDP-gluc)	1–5, 7–20, 22–24, 26–30	23, 24
17	8.8	$C_{27}H_{22}O_{19}$	649.0677	649.0679	0.31	605.06, 481.06, 479.04, 451.05, 425.01, 300.99(100), 298.99, 275.02, 273.00	Galloyl-HHDP-gluconate iso II	2, 18, 21	6
19	9.2	$C_{27}H_{22}O_{18}$	633.0733	633.0734	0.16	463.05, 300.99 (100), 275.02, 249.04, 169.01, 125.02	HHDP galloyl glucose Iso I	2–21, 26	6, 22
20	9.3	$C_{27}H_{22}O_{19}$	649.0677	649.0677	0.00	605.08, 481.06, 479.04, 451.05, 425.01, 300.99 (100), 298.99, 275.02, 273.00	Galloyl-HHDP-gluconate Iso III	2, 18, 21	6
21	9.5	$C_{41}H_{28}O_{27}$	951.0745	951.0749	0.42	907.08, 605.08, 450.99, 425.02 (100), 399.04, 301.00, 275.02	Trisgalloyl HHDP glucose Iso I	2–5	23
27	11	$C_{27}H_{22}O_{18}$	633.0733	633.0734	0.16	463.05, 300.99 (100), 275.02, 249.04, 169.01, 125.02	HHDP galloyl glucose Iso II	2–4, 7–22, 28	18, 22

(Continued)

TABLE 2.1 (Continued) HPLC-ESI-QTOF-MS and CID-MS/MS data of phytoconstituents from *Terminalia* species

PEAK NO.	RT (IN MIN)	MOLECULAR FORMULA	[M-H]- (CALC)	[M-H]- (OBS)	ERROR (ΔPPM)	MS/MS IN NEGATIVE ION MODE	IDENTIFICATION	*DETECTION IN SAMPLES	**REF.
28	11.4	$C_{34}H_{24}O_{22}$	783.0686	783.0681	−0.64	631.05, 450.99 (100)	Terflavin B	1–5, 7–20, 22–30	22
32	11.9	$C_{41}H_{26}O_{26}$	933.064	933.0636	−0.43	915.05, 889.07, 871.06, 631.06, 613.04, 587.06, 569.06, 551.04, 467.02, 441.04, 425.01, 300.99	Castalagin	1–5, 7–11, 13–16, 18–21, 23–30	24
35	12.5	$C_{34}H_{22}O_{22}$	781.053	781.0528	−0.26	721.03, 600.87 (100), 575.01, 448.98, 392.99, 298.98, 273.00, 271.99, 245.01	Punicalin α	1–20, 22	22
36	12.5	$C_{41}H_{26}O_{26}$	933.064	933.0643	0.32	781.05, 721.03, 631.06, 600.99, 450.99, 300.99, 125.03	2-O-galloylpunicalin Iso I	1–21, 23–30	–
38	13	$C_{48}H_{30}O_{30}$	1085.075	1085.075	−0.28	933.06, 783.07, 631.05, 600.99, 450.99(100), 425.01, 300.99	Terflavin A	2, 7–20, 22	–
40	13.1	$C_{17}H_{12}O_8$	343.0459	343.0451	−2.33	328.02, 312.99(100), 297.97, 285.00, 269.98	3, 3′, 4′-tri-O-methylellagic acid	1–5, 14, 15, 26, 29	–
41	13.4	$C_{34}H_{24}O_{22}$	783.0686	783.0686	0.00	481.06, 300.99(100), 275.02	Pedunculagin Iso II	1–5, 7–20, 22–30	23, 24
43	13.5	$C_{21}H_{10}O_{13}$	469.0049	469.0049	0.00	425.02(100), 407.01, 299.99	(S)-Flavogallonic acid	1–30	19

(Continued)

TABLE 2.1 (Continued)   HPLC-ESI-QTOF-MS and CID-MS/MS data of phytoconstituents from *Terminalia* species

PEAK NO.	RT (IN MIN)	MOLECULAR FORMULA	[M-H]⁻ (CALC)	[M-H]⁻ (OBS)	ERROR (ΔPPM)	MS/MS IN NEGATIVE ION MODE	IDENTIFICATION	*DETECTION IN SAMPLES	**REF.
47	13.9	$C_{41}H_{26}O_{26}$	933.064	933.0643	0.32	781.05, 721.03, 631.06, 600.99, 450.99, 425.01, 300.99, 275.02, 125.02	3-O-galloylpunicalin Iso II	1-5, 8-12, 14-25, 27-30	—
51	15	$C_{41}H_{28}O_{27}$	951.074	951.0749	0.95	907.09, 783.07(100), 764.05, 745.03, 605.08, 481.06, 300.99	Trisgalloyl HHDP glucose Iso II	2-5, 7-10, 15-18, 28-30	23
55	15.8	$C_{41}H_{28}O_{27}$	951.074	951.0748	0.84	907.09(100), 783.07, 745.03, 605.08, 481.06, 425.02, 300.99,	Trisgalloyl HHDP glucose Iso III	2-5, 7, 9, 10, 14-18	23
59	16.2	$C_{68}H_{48}O_{44}$	1567.146	1567.146	−0.13	783.07(100), 481.06, 300.99	Digalloyl triHHDP-diglucose	2	6
61	16.3	$C_{27}H_{22}O_{18}$	633.0733	633.0732	−0.16	463.05, 300.99(100), 275.02, 249.04, 169.01, 125.02	HHDP galloyl glucose Iso III	2-5, 7-25	18, 22
62	16.4	$C_{34}H_{24}O_{22}$	783.0686	783.0688	0.26	481.06, 300.99(100), 275.02	Pedunculagin Iso III	1-5, 7-20, 22-30	23, 24
67	17.3	$C_{34}H_{26}O_{22}$	785.0843	785.0842	−0.13	633.07, 615.06, 483.07, 419.06, 313.05, 300.99(100), 275.02, 249.04, 169.01	Tellimagrandin I	2-20, 22-24, 27-30	18, 22
71	17.9	$C_{27}H_{22}O_{18}$	633.0733	633.0735	0.32	463.05, 300.99 (100), 275.02, 249.04, 169.01, 125.02	HHDP galloyl glucose Iso IV	2-22, 26, 30	18, 22

(Continued)

TABLE 2.1 (Continued) HPLC-ESI-QTOF-MS and CID-MS/MS data of phytoconstituents from *Terminalia* species

PEAK NO.	RT (IN MIN)	MOLECULAR FORMULA	[M–H]⁻ (CALC)	[M–H]⁻ (OBS)	ERROR (ΔPPM)	MS/MS IN NEGATIVE ION MODE	IDENTIFICATION	*DETECTION IN SAMPLES	**REF.
81	19.3	$C_{20}H_{16}O_{13}$	463.0518	463.0517	−0.22	300.99(100)	Ellagic acid-hexoside	2–11, 14–21, 23–30	1, 18
85	19.9	$C_{34}H_{26}O_{22}$	785.0843	785.0849	0.76	633.07, 615.06, 483.07, 419.06, 313.05, 300.99(100), 275.02, 249.04, 169.01	DigallolylHHDPglucose	2–20, 22–24, 27–30	18
86	20.1	$C_{48}H_{28}O_{30}$	1083.059	1083.061	1.48	781.05, 600.99, 595.01, 450.99, 300.99	Punicalagin	2–11, 13–20, 22–25	22
87	20.2	$C_{41}H_{28}O_{27}$	951.0745	951.0748	0.32	907.09(100), 783.07, 764.05, 745.03, 605.08, 481.06, 425.02, 300.99, 275.02	Trisgalloyl HHDP glucose Iso IV	2, 17	23
94	21.1	$C_{41}H_{26}O_{26}$	933.064	933.0652	1.29	915.05, 889.07, 871.06, 631.06, 613.04, 587.06, 441.04, 425.01, 300.99	Vescalagin	1–5, 8–11, 14, 15, 17–20	24
101	21.5	$C_{20}H_{16}O_{12}$	447.0569	447.0569	0.00	300.99 (100), 299.99,	Ellagic acid-rhamnopyranoside Iso I	1–30	6
102	21.6	$C_{19}H_{14}O_{12}$	433.0412	433.0412	0.00	300.99(100), 299.99	Ellagic acid-pentoside	1–19, 21–30	18
107	22.4	$C_{41}H_{26}O_{26}$	933.064	933.0649	0.96	631.06, 481.06, 450.99(100), 425.02, 300.99, 275.02	Terflavin C	1–5, 8–11, 14–20	–

(Continued)

TABLE 2.1 (Continued)   HPLC-ESI-QTOF-MS and CID-MS/MS data of phytoconstituents from *Terminalia* species

PEAK NO.	RT (IN MIN)	MOLECULAR FORMULA	[M-H]- (CALC)	[M-H]- (OBS)	ERROR (ΔPPM)	MS/MS IN NEGATIVE ION MODE	IDENTIFICATION	*DETECTION IN SAMPLES	**REF.
108	22.6	$C_{22}H_{12}O_{13}$	483.0205	483.0205	0.00	450.99(100), 407.01, 299.99	Methyl (S)-flavogallonate	2, 4, 9, 10, 14, 15	22
110	22.7	$C_{20}H_{16}O_{12}$	447.0569	447.0569	0.00	300.99, 299.99(100)	Ellagic acid-rhamnopyranoside Iso II	1–30	6
115	23.3	$C_{13}H_8O_7$	275.0197	275.0198	0.36	257.01, 247.02, 229.01 (100), 219.03, 203.03, 191.03, 185.02, 173.02, 158.03, 145.03, 129.04	3, 4, 8, 9, 10-Pentahydroxy dibenzo [b, d] pyran-6-one	2, 8–10, 12, 15–17, 21	22
116	23.4	$C_{28}H_{10}O_{16}$	600.9896	600.9899	0.50	582.98, 300.99, 298.98, 270.99 (100), 242.99	Gallagic acid	1, 2, 4, 6–17, 27–30	22
117	23.5	$C_{41}H_{28}O_{26}$	935.0796	935.08	0.43	917.07, 873.08, 855.07, 783.07, 721.07, 633.07, 573.06, 553.06, 419.06, 383.04, 365.03, 349.04, 343.01, 329.03, 299.02 (100), 291.02, 275.02, 249.04	Casuarinin	2–5, 10	23
118	26.2	$C_{41}H_{28}O_{26}$	935.0796	935.08	0.43	917.07, 873.08, 783.07, 659.05, 633.07, 615.06, 571.07, 481.06, 419.06, 365.03, 329.03, 299.02, 275.02 (100), 249.04	Casuarictin	2–5, 10	

(Continued)

TABLE 2.1 (Continued) HPLC-ESI-QTOF-MS and CID-MS/MS data of phytoconstituents from *Terminalia* species

PEAK NO.	RT (IN MIN)	MOLECULAR FORMULA	[M–H]⁻ (CALC)	[M–H]⁻ (OBS)	ERROR (ΔPPM)	MS/MS IN NEGATIVE ION MODE	IDENTIFICATION	*DETECTION IN SAMPLES	**REF.
124	24.8	$C_{28}H_{24}O_{16}$	615.0992	615.0996	0.65	463.08 (100), 313.05, 301.03, 169.01	Dehydro-galloyl-HHDP-hexoside	1, 2, 16, 18, 25–28	6
127	24.9	$C_{41}H_{30}O_{26}$	937.0953	937.0957	0.43	785.08, 767.07, 599.07, 465.07, 419.06, 313.05, 300.99(100), 275.02, 169.01	Tellimagrandin II	2–4, 7, 9–12, 14–17, 22, 26–30	18
140	27.7	$C_{20}H_{16}O_{12}$	447.0569	447.0568	−0.22	315.02 (100), 299.99	3′-O-methyl-4-O-xylopyranosyl ellagic acid	1–30	22
152	30.4	$C_{21}H_{18}O_{12}$	461.0725	461.0725	0.00	315.01 (100), 299.99	3′-O-methyl-4-O-rhamnopyranosyl ellagic acid	1–30	–
153	30.5	$C_{27}H_{20}O_{16}$	599.0679	599.0678	−0.17	447.05, 300.99(100), 169.01	4-O-(4″-O-galloyl-rhamnopyranosyl) ellagic acid	1, 2, 6, 8–10, 16–25, 28	22
156	34	$C_{15}H_8O_8$	315.0146	315.0146	0.00	299.98(100)	3- O-Methyl ellagic acid	1–30	22
161	39.4	$C_{34}H_{24}O_{20}$	751.0788	751.0788	0.00	599.07, 447.05, 300.99(100), 169.01	4- O-(3″, 4″-di-O-galloyl-rhamnopyranosyl) ellagic acid	1–12, 14–28	22

(Continued)

TABLE 2.1 (Continued) HPLC-ESI-QTOF-MS and CID-MS/MS data of phytoconstituents from *Terminalia* species

PEAK NO.	RT (IN MIN)	MOLECULAR FORMULA	[M-H]⁻ (CALC)	[M-H]⁻ (OBS)	ERROR (ΔPPM)	MS/MS IN NEGATIVE ION MODE	IDENTIFICATION	*DETECTION IN SAMPLES	**REF.
166	40.6	$C_{16}H_{10}O_8$	329.0303	329.0303	0.00	314.01 (100), 298.98, 270.99	3, 3'-di-O-methyl ellagic acid	1–30	22
168	43.4	$C_{25}H_{22}O_{14}$	545.0937	545.0938	0.18	485.07, 470.05, 315.02, 314.00 (100), 298.98, 59.01	4'-O-methyl ellagic acid 3-(2'', 3''-di-O-acetyl)-rhamnoside	4, 6	–
Identification of proanthocyanidins									
4	6.2	$C_{45}H_{38}O_{21}$	913.1833	913.1837	0.44	787.15, 745.14, 727.13, 709.12, 619.11, 609.13, 607.11, 577.10, 559.09, 541.08, 483.09, 441.08, 429.08, 423.07, 391.05, 315.05, 305.07 (100), 303.05, 261.04, 243.03, 177.02, 125.03	(Epi)gallocatechin-(epi) gallocatechin-(epi) gallocatechin Iso I	3–5, 11, 14, 21, 26	17, 25
9	7.6	$C_{30}H_{26}O_{14}$	609.125	609.1249	−0.16	483.09, 441.08, 423.07 (100), 397.10, 355.08, 305.07, 303.05, 297.04, 283.02, 273.04, 261.07, 243.04, 219.07, 177.09, 137.03, 125.03	(Epi)gallocatechin-(epi) gallocatechin Iso I	1, 3–5, 11, 13–21, 25–28, 30	17, 25

(Continued)

TABLE 2.1 (Continued) HPLC-ESI-QTOF-MS and CID-MS/MS data of phytoconstituents from *Terminalia* species

PEAK NO.	RT (IN MIN)	MOLECULAR FORMULA	[M-H]⁻ (CALC)	[M-H]⁻ (OBS)	ERROR (ΔPPM)	MS/MS IN NEGATIVE ION MODE	IDENTIFICATION	*DETECTION IN SAMPLES	**REF.
10	7.9	$C_{30}H_{26}O_{14}$	609.125	609.1253	0.49	483.09, 441.08, 423.07(100), 305.07, 297.04, 283.02, 273.04, 261.07, 243.04, 219.07, 177.02, 137.02, 125.02	(Epi)gallocatechin-(epi) gallocatechin Iso II	1, 3–5, 11, 13–21, 25–28, 30	17, 25
16	8.8	$C_{45}H_{38}O_{21}$	913.1833	913.1836	0.33	787.15, 745.14, 727.13, 609.13, 607.11, 591.13), 577.10, 559.09, 541.07, 483.09, 441.08, 423.07, 305.07 (100), 303.05, 261.04, 243.03, 125.02	(Epi)gallocatechin-(epi) gallo catechin-(epi) gallocatechin Iso II	3–5, 11, 14, 21, 26	17, 25
23	9.8	$C_{45}H_{38}O_{21}$	913.1833	913.184	0.77	787.15, 745.14, 727.13, 609.13, 607.11, 577.10, 559.09, 541.07, 483.09, 441.08, 429.08, 423.07 (100), 391.05, 357.06, 347.07, 315.05, 305.07, 303.05, 273.04, 261.04, 243.03, 177.02, 125.02	(Epi)gallocatechin-(epi) gallo catechin-(epi) gallocatechin Iso III	3–5, 11, 14, 21, 26	17, 25
24	9.9	$C_{30}H_{26}O_{14}$	609.125	609.1248	−0.33	441.08, 423.07 (100), 305.07, 297.04, 283.02, 273.04, 261.07, 243.04, 219.07, 177.02, 137.02, 125.02	(Epi)gallocatechin-(epi) gallocatechin Iso III	1–5, 11, 13–21, 25–28, 30	17, 25

(Continued)

TABLE 2.1 (Continued)　HPLC-ESI-QTOF-MS and CID-MS/MS data of phytoconstituents from *Terminalia* species

PEAK NO.	RT (IN MIN)	MOLECULAR FORMULA	[M-H]⁻ (CALC)	[M-H]⁻ (OBS)	ERROR (ΔPPM)	MS/MS IN NEGATIVE ION MODE	IDENTIFICATION	*DETECTION IN SAMPLES	**REF.
26	10.1	$C_{30}H_{26}O_{13}$	593.1301	593.1305	0.67	467.11, 425.09, 407.08 (100), 381.09, 339.09, 303.05, 289.07, 273.08, 245.08, 177.02, 125.02	(Epi)gallocatechin-(epi) catechin Iso I	1–5, 11, 14, 16, 19, 21, 25–27, 30	17, 25
29	11.8	$C_{45}H_{38}O_{20}$	897.1878	897.1889	1.23	771.16, 729.14, 711.13, 593.13, 591.12, 559.08, 541.07, 467.09, 465.08, 447.07, 441.08, 429.08, 423.07, 405.06, 391.04, 305.06, 303.05, 301.03, 287.05, 261.04, 243.03, 177.02, 137.02, 125.02 (100)	(Epi)gallocatechin-(epi) catechin-(epi) gallocatechin Iso I	1, 3–5, 11, 16, 14, 19, 21, 26	17, 25
30	11.8	$C_{45}H_{38}O_{21}$	913.1833	913.1835	0.22	787.15, 745.14, 727.13, 709.12, 619.11, 609.13, 607.11, 577.09, 559.09, 541.08, 483.09, 441.08, 429.08, 423.07, 391.05, 315.05, 305.07 (100), 303.05, 261.04, 243.03, 177.02, 125.02	(Epi)gallocatechin-(epi) gallocatechin-(epi) gallocatechin Iso IV	3–5, 11, 14, 21, 26	17, 25

(Continued)

TABLE 2.1 (Continued) HPLC-ESI-QTOF-MS and CID-MS/MS data of phytoconstituents from *Terminalia* species

PEAK NO.	RT (IN MIN)	MOLECULAR FORMULA	[M-H]⁻ (CALC)	[M-H]⁻ (OBS)	ERROR (ΔPPM)	MS/MS IN NEGATIVE ION MODE	IDENTIFICATION	*DETECTION IN SAMPLES	**REF.
34	12.3	$C_{30}H_{26}O_{13}$	593.1301	593.1301	0	467.10, 423.07, 355.08, 305.07 (100), 289.07, 261.08, 219.07, 125.02	(Epi)catechin-(epi) gallocatechin Iso I	1–5, 11, 14, 16, 19, 21, 25, 26, 30	17, 25
37	12.8	$C_{45}H_{38}O_{20}$	897.1878	897.1877	−0.11	771.16, 729.14, 711.14, 607.11, 593.13, 559.08, 543.09, 541.07, 525.08, 505.08, 467.09, 441.08, 439.07, 429.08, 425.08, 423.07, 421.05, 413.08, 407.08, 391.05, 305.06, 303.05, 289.07, 261.04, 243.03, 177.02, 125.02 (100)	(Epi)gallocatechin-(epi) gallocatechin-(epi) catechin	1, 3–5, 11, 14, 16, 19, 21, 26	17, 25
44	13.6	$C_{30}H_{26}O_{13}$	593.1301	593.1303	0.34	425.09, 407.08(100), 381.10, 339.09, 289.07, 273.04, 245.08, 137.02, 125.02	(Epi)gallocatechin-(epi) catechin Iso II	1–5, 11, 16, 19, 21, 25, 26, 30	17, 25
45	13.6	$C_{37}H_{30}O_{18}$	761.1359	761.136	0.13	609.12, 591.11, 573.10, 483.08, 465.08, 453.08, 441.08, 423.07, 355.08, 305.07 (100), 303.05, 285.04, 273.04, 177.02, 137.02 (6), 125.02	3-O-galloyl-(epi) gallocatechin- (epi) gallocatechin Iso I	1, 3–5, 11, 14–16, 21, 26	17, 25

(Continued)

TABLE 2.1 (*Continued*)   HPLC-ESI-QTOF-MS and CID-MS/MS data of phytoconstituents from *Terminalia* species

PEAK NO.	RT (IN MIN)	MOLECULAR FORMULA	[M-H]⁻ (CALC)	[M-H]⁻ (OBS)	ERROR (ΔPPM)	MS/MS IN NEGATIVE ION MODE	IDENTIFICATION	*DETECTION IN SAMPLES	**REF.
46	13.9	$C_{45}H_{38}O_{20}$	897.1878	897.1878	0	771.16, 729.14, 711.14, 593.13, 591.12, 559.08, 541.07, 467.09, 465.08, 447.07, 441.08, 429.08, 423.07, 405.06, 391.04, 305.06, 303.05, 301.03, 287.05, 261.04, 243.03, 177.02, 137.02, 125.02(100)	(Epi)gallocatechin-(epi) catechin-(epi) gallocatechin Iso II	1, 3–5, 6, 7, 11, 14, 19, 22, 26	17, 25
48	14.2	$C_{30}H_{26}O_{13}$	593.1301	593.1303	0.34	467.10, 423.07, 355.08, 305.07(100), 289.07, 261.08, 245.09, 219.07, 125.03	(Epi)catechin-(epi) gallocatechin Iso II	1–5, 11, 14, 16, 21, 26, 27, 30	17, 25
49	14.4	$C_{30}H_{26}O_{14}$	609.125	609.1255	0.82	441.08, 423.07(100), 305.07, 297.04, 283.02, 273.04, 261.07, 243.04, 219.07, 177.02, 137.02, 125.02	(Epi)gallocatechin-(epi) gallocatechin Iso IV	1, 3–5, 11, 13–21, 25, 26, 28, 30	17, 25

(*Continued*)

TABLE 2.1 (Continued) HPLC-ESI-QTOF-MS and CID-MS/MS data of phytoconstituents from *Terminalia* species

PEAK NO.	RT (IN MIN)	MOLECULAR FORMULA	[M-H]⁻ (CALC)	[M-H]⁻ (OBS)	ERROR (ΔPPM)	MS/MS IN NEGATIVE ION MODE	IDENTIFICATION	*DETECTION IN SAMPLES	**REF.
50	14.8	$C_{45}H_{38}O_{19}$	881.1935	881.1931	−0.45	755.16, 729.14, 713.16, 711.13, 695.14, 593.13, 591.12, 577.13, 561.11, 545.10, 543.10, 541.09, 525.08, 467.10, 425.09, 423.07, 413.09, 407.08, 405.03, 303.05, 301.03, 289.07 (100), 287.06, 125.02	(Epi)gallocatechin-(epi) catechin-(epi)catechin Iso I	1, 3–5, 11, 16, 21, 25, 26, 30	17, 25
52	15.3	$C_{45}H_{38}O_{19}$	881.1935	881.1937	0.23	755.16, 729.13, 713.15, 711.13, 593.13, 585.11, 575.12, 561.10, 543.10, 541.09, 525.08, 467.10, 465.08, 449.09, 441.09, 423.08 (100), 413.09, 407.08, 405.03, 305.07, 289.07, 287.06, 177.02, 125.02	(Epi)catechin-(epi) catechin-(epi) gallocatechin Iso I	1, 3–5, 11, 16, 21, 26, 27, 30	17, 25
53	15.4	$C_{30}H_{26}O_{12}$	577.1351	577.1352	0.17	451.10, 425.08, 407.07, 381.10, 339.09, 299.05, 289.07(100), 245.08, 205.05, 161.02, 137.02, 125.02, 109.03	(Epi)catechin-(epi) catechin Iso I	1–6, 14–16, 18–22, 24–27, 29, 30	17, 25

(Continued)

TABLE 2.1 (Continued) HPLC-ESI-QTOF-MS and CID-MS/MS data of phytoconstituents from *Terminalia* species

PEAK NO.	RT (IN MIN)	MOLECULAR FORMULA	[M-H]⁻ (CALC)	[M-H]⁻ (OBS)	ERROR (ΔPPM)	MS/MS IN NEGATIVE ION MODE	IDENTIFICATION	*DETECTION IN SAMPLES	**REF.
56	16	$C_{30}H_{26}O_{13}$	593.1301	593.1303	0.34	467.10, 423.07(100), 355.08, 305.07, 289.07, 261.08, 219.07, 125.02	(Epi)catechin-(epi) gallocatechin Iso III	1, 3, 11, 14, 16, 21, 25, 26, 30	17, 25
57	16.2	$C_{37}H_{30}O_{18}$	761.1359	761.136	0.13	609.12, 591.11, 573.10, 547.12, 483.09, 465.08, 453.08, 441.08, 423.07, 305.07 (100), 297.04, 285.04, 273.04, 261.05, 177.02, 137.02, 125.02	3-O-galloyl-(epi) gallocatechin- (epi) gallocatechin Iso II	1–5, 11, 14, 15, 26	17, 25
58	16.2	$C_{45}H_{38}O_{19}$	881.1935	881.1931	−0.45	755.16, 729.13, 713.15, 711.11, 593.13, 585.11, 575.12, 561.10, 543.10, 541.09, 525.08, 467.10, 465.08, 449.09, 441.09, 423.07, 413.09, 407.08, 405.03, 305.07, 289.07, 287.06(100), 177.02, 125.02	(Epi)catechin-(epi) catechin-(epi) gallocatechin Iso II	1, 3–5, 11, 16, 21, 25, 26	17, 25
60	16.3	$C_{37}H_{30}O_{18}$	761.1359	761.1351	−1.05	609.13, 591.11, 483.09, 465.08, 453.08, 423.07 (100), 305.07, 301.03, 287.06, 269.05, 261.04, 243.03, 177.02, 169.01, 161.02, 125.02	(Epi)gallocatechin-3'-O-galloyl-(epi) gallocatechin	1–5, 11, 12, 14–16, 21, 26	17, 25

(Continued)

TABLE 2.1 (Continued) HPLC-ESI-QTOF-MS and CID-MS/MS data of phytoconstituents from *Terminalia* species

PEAK NO.	RT (IN MIN)	MOLECULAR FORMULA	[M-H]⁻ (CALC)	[M-H]⁻ (OBS)	ERROR (ΔPPM)	MS/MS IN NEGATIVE ION MODE	IDENTIFICATION	*DETECTION IN SAMPLES	**REF.
63	16.6	$C_{45}H_{38}O_{19}$	881.1935	881.1931	−0.45	755.16, 729.13, 713.14, 711.13, 695.14, 593.13, 591.11, 585.04, 577.13, 543.10, 541.09, 525.08, 467.10, 465.08, 449.09, 441.09, 425.09, 423.07, 413.09, 407.08, 405.03, 303.05, 301.03, 289.07, 287.06, 125.02 (100)	(Epi)gallocatechin-(epi) catechin-(epi)catechin Iso II	1, 3–5, 16, 21, 25, 26	17, 25
66	17.3	$C_{30}H_{26}O_{12}$	577.1351	577.135	−0.17	451.10, 425.08, 407.07, 381.109, 339.10, 299.05, 289.07(100), 245.08, 205.05, 161.02, 137.02, 125.02, 109.03	(Epi)catechin-(epi) catechin Iso II	1–6, 14–16, 18–22, 24–27, 29, 30	17, 25
69	17.7	$C_{37}H_{30}O_{18}$	761.1359	761.1355	−0.53	609.12, 591.11, 573.10, 547.12, 483.09, 465.08, 453.08, 441.08, 423.07, 305.07 (100), 297.04, 285.04, 273.04, 261.05, 177.02, 137.02, 125.02	3-O-galloyl-(epi) gallocatechin- (epi) gallocatechin Iso III	3, 11, 15, 26	17, 25

(Continued)

TABLE 2.1 (Continued) HPLC-ESI-QTOF-MS and CID-MS/MS data of phytoconstituents from *Terminalia* species

PEAK NO.	RT (IN MIN)	MOLECULAR FORMULA	[M-H]⁻ (CALC)	[M-H]⁻ (OBS)	ERROR (ΔPPM)	MS/MS IN NEGATIVE ION MODE	IDENTIFICATION	*DETECTION IN SAMPLES	**REF.
70	17.8	$C_{45}H_{38}O_{19}$	881.1935	881.1938	0.34	755.16, 729.13, 713.15, 711.13, 593.13, 585.11, 575.12, 561.10, 543.10, 541.10, 525.08, 467.10, 465.08, 449.09, 441.09, 423.07, 413.09, 407.08, 405.03 305.07, 289.07, 287.07(100), 177.02, 125.02	(Epi)catechin-(epi) catechin-(epi) gallocatechin Iso III	1, 3–5, 11, 16, 21, 25, 26	17, 25
73	18.1	$C_{45}H_{38}O_{18}$	865.1985	865.1984	−0.12	739.17, 713.15, 695.14, 577.13, 575.11, 543.09, 451.10, 449.08, 425.08, 413.09, 407.07, 405.06, 289.07, 287.05, 243.03, 161.02, 125.02(100), 109.02	(Epi)catechin-(epi) catechin-(epi)catechin Iso I	1–5, 10–16, 21, 22, 24, 26, 27	17, 25
75	18.4	$C_{30}H_{24}O_{14}$	607.1093	607.1095	0.33	571.09, 481.08, 469.08, 463.07, 439.07, 421.05, 395.07, 377.07, 355.05, 343.05, 337.03, 315.05, 313.03, 303.05, 301.03, 297.09, 283.03, 273.04, 259.03, 215.03, 179.04(100), 167.03, 137.02, 125.03	(Epi)gallocatechin-A-(epi)gallocatechin Iso I	1–6, 9, 10, 18, 21, 26, 29, 30	17, 25

(Continued)

TABLE 2.1 (Continued)   HPLC-ESI-QTOF-MS and CID-MS/MS data of phytoconstituents from *Terminalia* species

PEAK NO.	RT (IN MIN)	MOLECULAR FORMULA	[M–H]⁻ (CALC)	[M–H]⁻ (OBS)	ERROR (ΔPPM)	MS/MS IN NEGATIVE ION MODE	IDENTIFICATION	*DETECTION IN SAMPLES	**REF.
76	18.5	$C_{30}H_{26}O_{13}$	593.1301	593.1299	−0.34	467.09, 423.07, 355.08, 305.07 (100), 289.07, 261.08, 219.07, 161.02, 137.02, 125.02	(Epi)catechin-(epi) gallocatechin Iso IV	1–5, 11, 14, 16, 19, 21, 25, 26, 30	17, 25
77	18.5	$C_{37}H_{30}O_{17}$	745.141	745.1411	0.13	593.13, 575.12, 557.11, 467.09, 449.09, 441.08, 423.07, 407.08, 397.09, 305.07 (100), 300.99, 289.07, 287.06, 273.04, 269.05, 249.04, 245.05, 219.06, 177.02, 169.02, 161.02, 137.02, 125.02	3-O-galloyl(epi) catechin-(epi) gallocatechin	1–5, 11, 14–16, 21, 25, 26	17, 25
78	18.5	$C_{45}H_{38}O_{18}$	865.1985	865.1998	1.5	739.17, 713.15, 695.14, 577.13, 575.11, 543.09, 451.10, 449.08, 425.08, 413.09, 407.07, 405.06, 289.07, 287.05, 243.03, 161.02, 125.02(100), 109.03	(Epi)catechin-(epi) catechin-(epi)catechin Iso II	1–5, 10–16, 21, 22, 24, 26, 27	17, 25
82	19.4	$C_{30}H_{26}O_{12}$	577.1351	577.1352	0.17	451.10, 425.08, 407.07, 381.10, 339.09, 299.05, 289.07 (100), 245.08, 205.05, 161.02, 137.02, 125.02, 109.03	(Epi)catechin-(epi) catechin Iso III	1–6, 14–16, 18–22, 24–27, 29, 30	17, 25

(Continued)

TABLE 2.1 (Continued)   HPLC-ESI-QTOF-MS and CID-MS/MS data of phytoconstituents from *Terminalia* species

PEAK NO.	RT (IN MIN)	MOLECULAR FORMULA	[M-H]⁻ (CALC)	[M-H]⁻ (OBS)	ERROR (ΔPPM)	MS/MS IN NEGATIVE ION MODE	IDENTIFICATION	*DETECTION IN SAMPLES	**REF.
83	19.5	$C_{37}H_{30}O_{17}$	745.141	745.1415	0.67	593.13, 575.13, 467.10, 423.07 (100), 405.06, 387.06, 305.07, 299.06, 287.06, 269.05, 261.03, 255.03, 245.05, 243.03, 229.01, 177.02, 169.02, 161.02, 125.02	(Epi)catechin-3′-O-galloyl-(epi) gallocatechin	1–5, 11, 14–16, 21, 25, 26	17, 25
88	20.4	$C_{37}H_{30}O_{17}$	745.141	745.1407	−0.4	593.13, 575.12, 467.10, 445.04, 423.07, 407.08 (100), 389.07, 305.07, 301.00, 289.07, 287.06, 275.02, 271.06, 177.02, 169.01, 125.02	(Epi)gallocatechin-3′-O-galloyl-(epi) catechin	1–5, 11, 14–16, 21, 25, 26	17, 25
90	20.8	$C_{30}H_{24}O_{14}$	607.1093	607.1095	0.33	571.09, 481.08, 469.08, 463.07, 439.07, 421.05, 395.07, 377.07, 355.05, 343.05, 337.03, 315.05, 313.03, 303.05, 301.03, 297.09, 283.03, 273.04, 259.03, 215.03, 179.04(100), 167.03, 137.02, 125.03	(Epi)gallocatechin-A-(epi)gallocatechin Iso II	1, 3–6, 9, 10, 18, 21, 26–30	17, 25

(Continued)

TABLE 2.1 (*Continued*) HPLC-ESI-QTOF-MS and CID-MS/MS data of phytoconstituents from *Terminalia* species

PEAK NO.	RT (IN MIN)	MOLECULAR FORMULA	[M-H]⁻ (CALC)	[M-H]⁻ (OBS)	ERROR (ΔPPM)	MS/MS IN NEGATIVE ION MODE	IDENTIFICATION	*DETECTION IN SAMPLES	**REF.
97	21.3	$C_{37}H_{30}O_{16}$	729.1461	729.146	−0.14	559.12, 451.10, 441.08, 407.08 (100), 289.07, 287.06, 271.06, 245.05, 169.01, 161.02, 125.02	3-O-Galloyl(epi) catechin-(epi)catechin Iso I	1–5, 9, 11, 13–16, 18–22, 24–27	17, 25
103	21.8	$C_{30}H_{26}O_{12}$	577.1351	577.1356	0.87	451.10, 425.08, 407.07, 381.10, 339.09, 299.05, 289.07(100), 245.08, 205.05, 161.02, 137.02, 125.02, 109.03	(Epi)catechin-(epi) catechin Iso IV	1–6, 14–16, 18–22, 24–27, 29, 30	17, 25
104	21.9	$C_{45}H_{38}O_{18}$	865.1985	865.1989	0.46	739.17, 713.15, 695.14, 577.13, 575.11, 543.09, 451.10, 449.08, 425.08, 413.09, 407.07, 405.06, 289.07, 287.05, 243.03, 161.02, 125.02 (100), 109.03	(Epi)catechin-(epi) catechin- (epi) catechin Iso III	1–5, 10–16, 21, 22, 24, 26, 27	17, 25
113	23.2	$C_{30}H_{26}O_{12}$	577.1351	577.1351	0	451.10, 425.09, 407.08, 381.09, 339.09, 299.09, 289.07 (100), 245.0, 205.05, 161.02, 137.02, 125.02	(Epi)catechin-(epi) catechin Iso V	1–6, 14–16, 18–22, 24–27, 29, 30	17, 25

(*Continued*)

TABLE 2.1 (Continued) HPLC-ESI-QTOF-MS and CID-MS/MS data of phytoconstituents from *Terminalia* species

PEAK NO.	RT (IN MIN)	MOLECULAR FORMULA	[M-H]⁻ (CALC)	[M-H]⁻ (OBS)	ERROR (ΔPPM)	MS/MS IN NEGATIVE ION MODE	IDENTIFICATION	*DETECTION IN SAMPLES	**REF.
114	23.2	$C_{37}H_{30}O_{16}$	729.1461	729.1459	−0.27	559.12, 451.10, 441.08, 407.08 (100), 289.07, 287.06, 271.06, (245.05, 169.01, 161.02, 125.02	(Epi)catechin-3'-O-galloyl-(epi)catechin Iso I	1–5, 9, 11, 13–16, 18–22, 24–27	17, 25
146	28.9	$C_{37}H_{30}O_{16}$	729.1461	729.1461	0	577.13, 559.12, 541.11, 451.10, 433.09, 425.09, 407.07, 381.09, 289.07 (100), 287.05, 269.04, 245.04, 169.01, 125.02	3-O-Galloyl(epi)catechin-(epi)catechin Iso II	1–5, 9, 13–16, 18–22, 24–27	17, 25
155	31.5	$C_{37}H_{30}O_{16}$	729.1461	729.146	−0.14	577.12, 559.12, 451.09, 541.11, 441.08, 433.09, 425.09, 407.07 (100), 381.09, 289.07, 271.06, 245.04, 169.01, 137.02, 125.02	(Epi)catechin-3'-O-galloyl-(epi)catechin Iso II	1–5, 9, 13–16, 18–22, 24–27	17, 25

*1= *T. arjuna* bark; 2= *T. arjuna* fruit; 3= *T. arjuna* leaf; 4= *T. arjuna* stem; 5= *T. arjuna* root; 6= *T. bellirica* bark; 7= *T. bellirica* fruit; 8= *T. bellirica* leaf; 9= *T. bellirica* stem; 10= *T. bellirica* root; 11= *T. catappa* bark; 12= *T. catappa* fruit; 13= *T. catappa* leaf; 14= *T. catappa* stem; 15= *T. catappa* root; 16= *T. chebula* bark; 17= *T. chebula* fruit; 18= *T. chebula* leaf; 19= *T. chebula* stem; 20= *T. chebula* root; 21= *T. elliptica* bark; 22= *T. elliptica* fruit; 23= *T. elliptica* leaf; 24= *T. elliptica* stem; 25= *T. elliptica* root; 26= *T. paniculata* bark; 27= *T. paniculata* fruit; 28= *T. paniculata* leaf; 29= *T. paniculata* stem; 30= *T. paniculata* root.
ᵃ Confirmed with the reference standard.

**(1). Taamalli et al. 2014; (2). MassBank Record: PR100485; (3). MassBank Record: KOX00321; (4). MassBank Record: PT202000; (5). Fang et al. 2002; (6). Mena et al. 2012; (7) Sanchez-Rabaneda et al. 2003; (8). MassBank Record: FIO00624; (9). MassBank Record: BML00729; (10). MassBank Record: PR100984; (11). MassBank Record: BML00601; (12). MassBank Record: BML82316; (13). MassBank Record: FIO00705; (14). Downeya and Rochfort 2008; (15). Wan et al. 2013; (16). Zhou et al. 2012; (17). Jaiswal et al. 2012; (18). Yang et al. 2012; (19). Bystrom et al. 2008; (20). Fabre et al. 2001; (21). MassBank Record: BML01698; (22). Pfundstein et al. 2010; (23). Barry et al. 2001; (24). Zywicki et al. 2002 (25). Callemien and Collin 2008.

2.5.3 Identification of Flavonoids

Flavonoids are an important group of phenolic compounds widely distributed in *Terminalia* species (Manipal et al. 2017). Thirty-five flavonoids were identified and characterized, out of which 19 compounds (**80, 93, 98–100, 121, 125, 131, 134, 154, 159, 163–165, 169, 171–173, 178**) were positively identified by comparing their RT, MS and MS/MS pattern with those of authentic reference standards (Table 2.1). Two compounds **39** and **54** at *m/z* 305.0667 generated similar fragment ions at *m/z* 167.03 and 137.02 by RDA fragmentation as expected for (+)-gallocatechin or (−)-epigallocatechin (Yuzuak et al. 2018). As epigallocatechin is more hydrophobic than gallocatechin, in reversed-phase HPLC, the latter elutes first and, therefore, compounds **39** and **54** were identified as (+)-gallocatechin and (−)-epigallocatechin, respectively. Compound **120** at *m/z* 477.0674 was tentatively identified as 3-*O*-methylquercetin-7-*O*-glucopyranoside on the basis of the MS/MS generated fragment ions at *m/z* 315.01 and 299.99 corresponding to [M-H]⁻ and [M-CH₃]⁻ of methyl quercetin, respectively. Compounds **145** and **174** at *m/z* 599.1042 and 313.0718 were tentatively identified as kaempferol-*O*-galloyl-hexoside and dihydroxydimethoxyflavone, respectively with literature support (Taamalli et al. 2014, MassBank Record: BML01698). Compound **157** at *m/z* 317.0303 was identified as myricetin with fragment ions at *m/z* 151.00 by RDA fragmentation (Lin et al. 2012). Similarly, five myricetin derivatives (**96, 109, 111, 122** and **132**) were also tentatively identified. Six quercetin derivatives (**95, 121, 126, 133, 136** and **149**) were tentatively identified and their fragment ions were matched with previously reported literature (Table 2.1). Fragment ions at *m/z* 301.03 and 300.03 were generated for all the quercetin derivatives corresponding to the quercetin moiety.

2.5.4 Identification of GA Derivatives

Twenty-seven compounds were tentatively identified as GA derivatives. They were esters of GA and polyol, usually glucose, flavonoids and shikimic acid. Thirteen gallotannins were identified as monogalloyl glucose (**5, 11**), digalloyl glucose (**18, 22**), trigalloyl glucose (**65, 79**), tetragalloyl glucose (**106, 119, 123, 135, 137**), pentagalloyl glucose (**141, 151**) and their isomers. They showed characteristic fragment ions in their product ion spectra by consecutive elimination of galloyl and gallate moieties (Chang et al. 2019). Pentagalloyl glucose (*m/z* 939.1109) produced consistent losses of galloyl (G, Δm = 152) moieties to tetragalloyl glucose (*m/z* 787.0999), trigalloyl glucose (*m/z* 635.0884), digalloyl glucose (*m/z* 483.078) and monogalloyl glucose (*m/z* 331.0671). They produced fragment ions corresponding to the loss of a GA moiety (GOH, Δm = 170).

Fragment ions at m/z 271.05, 169.01 and 125.02 corresponded to glucose cross-ring fragmentation, and deprotonation and decarboxylation of the GA moieties, respectively. Compounds **31** and **33** produced similar ions (m/z 325.0565) and MS/MS fragmentation patterns indicated these to be positional isomers. Fragment ions were observed at m/z 169.02 and 125.03 due to deprotonated and decarboxylated ions of the GA moieties, respectively. A fragment ion at m/z 137.02 was observed by the loss of GA (GOH, $\Delta m = 170$) and water (H_2O, $\Delta m = 18$) molecules from [M-H]$^-$. The m/z 137.02 ion underwent CO_2 loss to generate an ion at m/z 93.04. A similar fragment ion at m/z 111.05 was observed by the loss of GA (GOH, $\Delta m = 170$) and CO_2 ($\Delta m = 44$). Hence, compounds **31** and **33** were tentatively identified as galloyl shikimic acid isomers. Digalloyl shikimic acid (m/z 477.0675) showed loss of galloyl moiety (G, $\Delta m = 152$) resulting in galloyl shikimic acid (m/z 325.06), whereas trigalloyl shikimic acid (m/z 629.0784) exhibited consecutive losses of galloyl moieties (G, $\Delta m = 152$) to digalloyl shikimic acid (m/z 477.07) and galloyl shikimic acid (m/z 325.06) (Chang et al. 2019). Hence, compounds **84** and **89** were identified as the [M-H]$^-$ ion of isomeric digalloyl derivatives of shikimic acid and **130** was identified as the trigalloyl derivative of shikimic acid.

Compound **68** at m/z 453.1038 was identified as *3*-methoxy-*4*-hydroxyphenol-*1*-*O*-*β*-*D*-(*6'*-*O*-galloyl)-glucoside based on its MS/MS product ions. Compounds **91** at m/z 183.0299 and **144** at m/z 197.0455 both producing fragment ion at m/z 124.02 corresponding to (M-H-CH$_3$/C$_2$H$_5$-CO$_2$)$^-$ were identified as methyl and ethyl gallate, respectively. Five compounds **92, 105, 139, 142** and **147** were tentatively identified as (+)-gallocatechin-*3*-*O*-gallate, (−)-epigallocatechin-*3*-*O*-gallate, myricetin-*3*-*O*-gallate, (+)-catechin-*3*-*O*-gallate and (−)-epicatechin-*3*-*O*-gallate, respectively. The exact mass, molecular formula and fragment ions of these molecules at m/z 169.01 and 125.02 due to gallate ion and [gallate-CO$_2$]$^-$ ion, respectively, together with their retention behavior supported the identification. Compound **143** at m/z 583.1093 generated fragment ions at m/z 431.09 by the loss of a galloyl group, m/z 341.06 and 311.05 by further fragmentation of the hexose ring, m/z 269.05 by the loss of a galloyl hexoside group, m/z 169.01 due to gallate ion and m/z 125.02 due to [gallate-CO$_2$]$^-$ ion indicating the presence of vitexin-2″-*O*-gallate.

2.5.5 Identification of EA Derivatives

EA derivatives are hydrolyzable tannins since they are esters of hexahydroxy-diphenic acid (HHDP: 6,6′-dicarbonyl-2,2′, 3,3′, 4,4′-hexahydroxybiphenyl) and a polyol, usually glucose, and in some cases GA. They were distinguished by their characteristic fragment ion spectra yielding sequential losses of galloyl (m/z 152), gallate or GA (m/z 170) and HHDP (m/z 301) residues (Mena et al. 2012).

The general substitution patterns of HHDP and galloyl groups for 50 EA derivatives were determined from the exact mass information and characteristic MS/MS fragmentation pattern. If molecular weights of hydrolyzable tannins differ by two units, it can be due to the difference between an HHDP group and two galloyl groups. For example, coupling of two adjacent galloyl groups from tetra-galloyl-glucose (788) by intramolecular oxidation will result in tellimagrandin I (di-galloyl-HHDP-glucose, 786) and further coupling of the two remaining galloyl groups from tellimagrandin I will produce pedunculagin (di-HHDP-glucose, 784).

In the MS/MS spectrum of the [M-H]⁻ ion of compound **7** at m/z 481.0624 fragment ions were observed at m/z 421.04 due to cleavage of a glucose ring, m/z 300.99 corresponding to an HHDP residue and m/z 275.02 due to decarboxylation of the HHDP moiety. Hence, compound **7** was identified as HHDP-glucose. Similarly, three compounds **15, 41** and **62** at m/z 783.0686 were identified as diHHDP-glucose (pedunculagin) isomers, all of which produced fragment ions at m/z 481.06 corresponding to HHDP-glucose, m/z 300.99 corresponding to the HHDP residue, and m/z 275.02 by decarboxylation of the HHDP moiety in the MS/MS spectrum. For castalin and vescalin, both being known compounds in *Terminalia* species, two peaks were expected at m/z 631 but only one, **8**, was observed at m/z 631.0577. Therefore, compound **8** was tentatively identified as either castalin or vescalin. The MS/MS spectra of the two compounds **12** and **35** at m/z 781.053 were similar to those of punicalin isomers (α/β-isomer). In reverse phase HPLC, the β-isomer elutes before the α-isomer; therefore, compound **12** was tentatively identified as punicalin (punicalin β) and compound **35** as isopunicalin (α-isomer).

Galloyl-HHDP-gluconate, HHDP-galloyl-glucose, digalloyl-HHDP-glucose, trigalloyl-HHDP-glucose, trisgalloyl-HHDP-glucose and their isomers were tentatively identified by their relevant fragments. Compound **28** generated fragment ions at m/z 631.05 (loss of the galloyl residue) and 450.99 (loss of galloyl glucose) suggesting that **28** is terflavin B. Four compounds **19, 27, 61** and **71** at m/z 633.0733 were tentatively identified as HHDP galloyl glucose isomers. They generated fragment ions at m/z 463.05 due to the loss of GA and m/z 300.99 indicating the presence of the HHDP group. Similarly, four compounds **21, 51, 55** and **87** with [M-H]⁻ at m/z 951.0745, two compounds **67** and **85** at m/z 785.0843 and one compound **127** at m/z 937.0953 were identified as trigalloyl-HHDP-glucose isomers, di galloyl-HHDP-glucose (tellimagrandin I) isomers and trigalloyl-HHDP-glucose (tellimagrandin II), respectively. Five compounds **32, 36, 47, 94** and **107** showed [M-H]⁻ion at m/z 933.064. Two of them **32** and **94** generated similar MS/MS fragmentation patterns producing fragment ions at m/z 915.05 by loss of a water molecule, m/z 631.06 by loss of EA, m/z 613.04 by loss of a water molecule followed by EA and m/z 300.99 corresponding to deprotonated EA. They were identified as castalagin and vescalagin (stereoisomers). Two other compounds **36** and **47** at m/z 933.064 produced similar MS/MS fragmentation and were tentatively identified as isomers of 2-*O*-galloylpunicalin or 3-*O*-galloylpunicalin. Compound **107** was tentatively identified as terflavin C. It generated fragment ions at m/z 631.06 by loss of HHDP, 481.06 by loss of an ellagic gallate moiety and 450.99 and 300.99 corresponding to an ellagic gallate moiety and EA, respectively.

Compound **38** at m/z 1085.0749 generated fragment ions at m/z 933.06 by loss of a galloyl residue, 783.07 by loss of HHDP and 631.05 by loss of HHDP followed by the galloyl residue. Fragment ions at m/z 450.99 and 300.99 were due to an ellagic galloyl residue and GA residue, respectively. Therefore, compound **38** was tentatively identified as terflavin A. Compounds **40, 156** and **166** at m/z 343.0451, 315.0146 and 329.0303 respectively were identified as trimethyl (**40**), methyl (**156**) and dimethyl (**166**) derivatives of EA respectively as confirmed by their MS and MS/MS. Compound **43** generated fragment ions at m/z 425.02 and 407.01 by loss of CO_2 followed by loss of water. Similarly, its methyl derivative (**108**) yielded fragment ions at m/z 450.99 by loss of methanol. Therefore, compounds **43** and **108** were identified as (*S*)-flavogallonic acid and methyl (*S*)-flavogallonate, respectively. Their product ion spectra also matched with those previously reported (Bystrom et al. 2008, Pfundstein et al. 2010).

Compound **81** at 463.0518 was identified as ellagic acid hexoside, which produced a fragment ion at m/z 300.99 (EA). Compound **86** at m/z 1083.0593 was tentatively identified as punicalagin and its MS/MS pattern was also matched with previous literature (Pfundstein et al. 2010). Three compounds **101, 110** and **140** were observed at m/z 447.0569. In the MS/MS spectrum of compounds **101** and **110**, fragment ions at m/z 300.99 (loss of a rhamnopyranosyl residue) and m/z 299.99 (loss of H•) corresponded to an EA moiety (Mena et al. 2012). Therefore, compounds **101** and **110** were tentatively identified as ellagic acid rhamnopyranoside isomers. Compound **140** generated fragment ion (loss of pentose) at m/z 315.02 (*3'-O*-methyl ellagic acid) and 299.99 (loss of •CH_3). Therefore, compound **140** was tentatively identified as *3'-O*-methyl-*4-O*-xylopyranosyl ellagic acid. Peaks **102** and **115** at m/z 433.0412 and 275.0197 were tentatively identified as ellagic acid pentoxide and 3,4,8,9,10-pentahydro xydibenzo[b,d]pyran-6-one, respectively on the basis of their exact mass and fragmentation patterns. Compound **116** at m/z 600.9896 was tentatively identified as gallagic acid on the basis of our analytical data with literature support (Pfundstein et al. 2010).

Compound **124** (m/z 615.0996) was identified as dehydro-galloyl-HHDP-hexoside and confirmed with the product ion spectrum as given in the literature (Mena et al. 2012). Compound **152** generated fragment ions at m/z 315.01 by loss of a rhamnopyranosyl residue and m/z 299.99 by further loss of a methyl radical. Therefore, it was tentatively identified as *3'-O*- methyl-*4-O*-rhamnopyranosyl ellagic acid.

Compound **153** at m/z 599.0679 and **161** at 751.0788 were identified as monogalloyl and digalloyl derivatives of rhamnopyranosyl ellagic acid, respectively. Fragment ions at m/z 447.05 and 300.99 were generated due to loss of galloyl, galloyl rhamnopyranoside residues for compound **153** and fragment ions at m/z 599.08 and 300.99 were generated due to loss of galloyl and digalloyl rhamnopyranoside residues for compound **161**. Compound **168** at m/z 545.0937 produced fragment ions at m/z 485.07 due to loss of acetic acid, m/z 470.05 due to further loss of a methyl radical, m/z 315.02 due to methyl ellagic acid (loss of a diacetyl rhamnose residue) and 314.00 due to loss of H· from m/z 315.02. Therefore, compound **168** was tentatively identified as 4'-*O*-methylellagic acid 3-(2″, 3″-di-*O*-acetyl)-rhamnoside.

2.5.6 Identification of PAs

PAs are polymers (oligomers) of catechin and their ent-isomer. A total of 43 PAs including dimers, trimers and their galloyl derivatives have been identified in different parts of *Terminalia* species (Singh et al. 2018b). Heterocyclic ring fission (HRF), retro-Diels–Alder (RDA) fragmentation, quinone methide (QM) fragmentation and benzofuran formation (BFF) have been identified as the characteristic fragmentation pathways. HRF and RDA fragmentation provide information about the hydroxylation of the *B*-rings and bonds between two monomeric units while QM fragmentation defines two monomeric units, especially the base unit of PAs (Li and Deinzer 2007). Structure elucidation of tannins from *Terminalia* species has been reviewed (Chang et al. 2019).

2.5.6.1 Dimeric B and A-type PAs

Peaks **9, 10, 24** and **49** at m/z 609.1250 with similar MS/MS spectrum patterns were tentatively identified as dimeric *B*-type PAs with (epi)gallocatechin monomeric units. They produced fragment ions at m/z 441.08 and 423.07 by RDA fragmentation and further loss of a water molecule, while the ion at m/z 305.07 was obtained by QM fragmentation. Peaks **26, 34, 44, 48, 56** and **76** at m/z 593.1301 were tentatively assigned to dimeric *B*-type PAs with (epi) catechin and (epi)gallocatechin monomeric units. Two RDA fragmentations are possible in these compounds either on the top unit or on the base unit. The RDA fragmentation on the top unit produces a more stable ion due to larger *p-p* hyperconjugation, which is energetically more stable than the other one (Tala et al. 2013). Peaks **26** and **44** produced similar fragmentation patterns in the MS/MS spectrum at m/z 425.09 and 407.08 by RDA and further loss of water molecule indicating (epi)gallocatechin as the top unit (Figure 2.2).

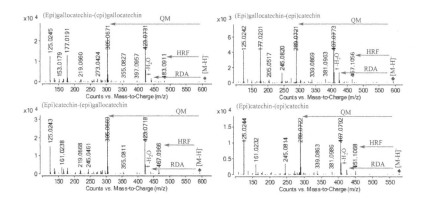

FIGURE 2.2 (–)-ESI-MS/MS spectra of *B*-type proanthocyanidin dimers.

Fragment ion at *m/z* 289.07 observed from QM fragmentation confirmed (epi) catechin as the base unit. Therefore, **26** and **44** were assigned to isomers of dimeric (epi)gallocatechin-(epi)catechin. Peaks **34, 48, 56** and **76** also showed similar fragmentation patterns in their MS/MS spectra and generated fragment ion at *m/z* 423.07 by RDA fragmentation followed by loss of water indicating the top unit as (epi)catechin. Fragment ions at *m/z* 467.10 and 305.08 generated by HRF and QM fragmentation, respectively, confirmed the base unit as (epi)gallocatechin. Thus, Peaks **34, 48, 56** and **76** are tentatively identified as isomers of (epi)catechin-(epi)gallocatechin (Figure 2.2).

Peaks **53, 66, 82, 103** and **113** at *m/z* 577.1359 produced fragment ions analogous to isomers **9, 10, 24** and **49** and could be assigned to dimeric (epi) catechin-(epi)catechin. Peaks **75** and **90** at *m/z* 607.1095 with similar MS/MS patterns were tentatively identified as *A*-type PAs with an (epi)gallocatechin monomeric unit. They produced fragment ions at *m/z* 481.08 by HRF, *m/z* 439.07 by RDA, *m/z* 421.05 by RDA followed by loss of water molecule, *m/z* 343.05 by a BFF, 303.05 by QM and 571.09 due to [M-H-2H$_2$O]$^-$. Peaks **45, 57, 60** and **69** at *m/z* 761.1359 were tentatively assigned to gallate of the dimeric PAs with (epi)gallocatechin-(epi)gallocatechin. There are two possibilities for the attachment of a galloyl residue, either it is attached to the C3 of the top unit or it is attached to C3′ of the base unit. If the attachment of the galloyl residue is to the C3′ of the base unit, then it will favor the loss of GA over the RDA fragmentation of the top unit (Tala et al. 2013). Peaks **45, 57, 60** and **69** produced fragment ions at *m/z* 609.12 and *m/z* 591.11 due to loss of a galloyl residue and GA, respectively. Fragment ion at *m/z* 305.07 was formed by QM fragmentation of ion [M-H-C$_7$H$_5$O$_4$]$^-$ (Figure 2.3). Peaks **45, 57** and **69** showed a characteristic RDA fragment at *m/z* 441.08, which was not detected in peak **60**. It indicated attachment of the galloyl residue with C3 of the top

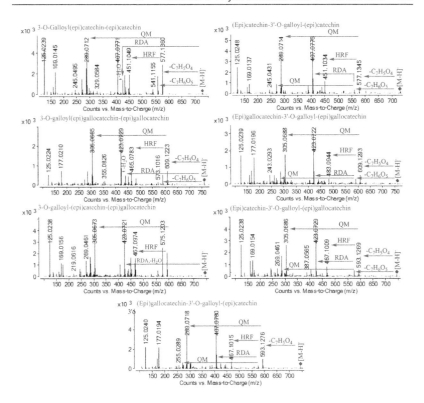

FIGURE 2.3 (–)-ESI-MS/MS spectra of galloyl derivatives of *B*-type proanthocyanidin dimers.

unit for peaks **45, 57** and **69**, whereas the attachment is with C3′ of the base unit for peak **60**. Hence, peaks **45, 57** and **69** were assigned to 3-*O*-galloyl-(epi)gallocatechin-(epi)gallocatechin and peak **60** to (epi)gallocatechin-3′-*O*-galloyl-(epi)gallocatechin.

Similarly, peaks **77, 83** and **88** at *m/z* 745.141 were tentatively assigned to gallate of the dimeric PAs with (epi)gallocatechin and (epi)catechin units. Peaks **77, 83** and **88** were identified as 3-*O*-galloyl-(epi)catechin-(epi)gallocatechin, (epi)catechin-3′-*O*-galloyl- (epi)gallocatechin and (epi)gallocatechin-3′-*O*-galloyl-(epi)catechin, respectively based on the characteristic fragment ions (Figure 2.3). Peaks **97, 114, 146** and **155** at *m/z* 729.1461 were tentatively assigned to gallate of the dimeric PAs with (epi) catechin monomeric units. These peaks also produced MS/MS fragment ions analogous to isomers **45, 57, 60** and **69** and could be readily assigned to isomers of 3-*O*-galloyl-(epi)catechin-(epi)catechin (**97** and **146**) and (epi)catechin-3′-*O*-galloyl-(epi)catechin (**114** and **155**).

2.5.6.2 Trimeric B-type PAs

Peaks **4, 16, 23** and **30** at *m/z* 913.1835 were tentatively identified as trimeric *B*-type PAs with (epi)gallocatechin monomeric units. They generated fragment ions at *m/z* 787.15 by loss of a galloyl residue (HRF fragmentation), *m/z* 745.14 by RDA fragmentation and *m/z* 727.13 by further loss of water. Fragment ions at *m/z* 607.11 and 305.07 were generated by QM. lower fragmentation. Similarly, fragment ions at *m/z* 609.13 and 303.05 were generated by QM upper fragmentation. Fragment ion at *m/z* 441.08 formed by RDA fragmentation of *m/z* 609.13 and further loss of water molecule gave fragment ion at *m/z* 423.07 (Figure 2.4). These PAs were assigned as isomers of trimeric (epi)gallocatechin-(epi)gallocatechin-(epi)gallocatechin.

Peaks **29, 37** and **46** at *m/z* 897.1878 were tentatively assigned to the trimeric PAs with one (epi)catechin and two (epi)gallocatechin monomeric units. Peaks **29** and **46** produced similar MS/MS spectra and produced fragment ions at *m/z* 771.16 by HRF fragmentation and *m/z* 467.09 by further HRF fragmentation, and *m/z* 729.14 by RDA fragmentation, which showed the upper unit as (epi)gallocatechin. Other fragment ions generated at *m/z* 305.06, 591.12 by QM lower fragmentation, and *m/z* 593.13, 303.05 by QM upper fragmentation have confirmed the middle unit as (epi)catechin and the bottom unit as (epi) gallocatechin. Fragment ions at *m/z* 441.08 by RDA fragmentation followed by loss of a water molecule at *m/z* 423.07 obtained from ion at *m/z* 593.13 in

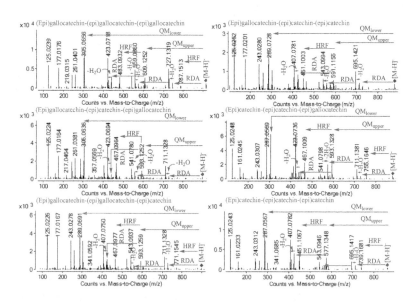

FIGURE 2.4 (–)-ESI-MS/MS spectra of *B*-type proanthocyanidin trimers.

high intensity confirmed our findings. Hence, peaks **29** and **46** were tentatively identified as (epi)gallocatechin-(epi)catechin-(epi)gallocatechin.

In the MS/MS spectrum of peak **37** at *m/z* 897.1878, fragment ions were generated at *m/z* 771.16 by loss of a galloyl residue (HRF fragmentation), *m/z* 467.09 by further HRF fragmentation, *m/z* 729.14 by RDA fragmentation and *m/z* 711.14 by further loss of water indicating the upper unit as (epi)gallocatechin. Other fragment ions generated at *m/z* 289.07 and 607.11 by QM lower fragmentation, and *m/z* 593.13 and 303.05 by QM upper fragmentation confirmed the middle unit as (epi)gallocatechin and the bottom unit as (epi)catechin. RDA fragmentation of the ion at *m/z* 593.13 leads to the ion at *m/z* 425.08 and 407.08 by further loss of water. Hence, peaks **37** was tentatively identified as (epi)gallocatechin-(epi)gallocatechin-(epi)catechin. Peaks **50, 52, 58, 63** and **70** at *m/z* 881.1935 were tentatively assigned to the trimeric PAs with two (epi)catechin monomeric units and one (epi)gallocatechin monomeric unit. They produced the MS/MS fragment ions analogously to isomers **29, 37** and **46** and could be readily assigned as isomers of (epi)gallocatechin-(epi)catechin-(epi)catechin (**50** and **63**) and (epi)catechin-(epi)catechin-(epi)gallocatechin (**52, 58** and **70**). Peaks **73, 78** and **104** detected at *m/z* 865.1985 were tentatively assigned to trimeric C-type PAs with (epi)catechin monomeric units. They produced MS/MS fragment ions analogous to isomers **4, 16, 23** and **30** and could be readily assigned to isomers of trimeric (epi)catechin-(epi)catechin- (epi)catechin trimer.

2.6 CONCLUSION

In summary, a simple, reliable and reproducible method for screening the phytochemicals of *Terminalia* species was developed using HPLC-ESI-QTOF-MS/MS. Different classes of compounds mainly polyphenols, such as acids, flavonoids, GA derivatives, EA derivatives and PAs have been identified and characterized by their RT, exact mass measurement, isotopic peak pattern, molecular formula and MS/MS fragmentation patterns. A total of 179 compounds including 25 acids, 35 flavonoids, 27 GA derivatives, 49 EA derivatives and 43 PAs were tentatively identified from various parts of the six *Terminalia* species. The developed method is fast and accurate and can be easily used for the identification and characterization of phytoconstituents in *Terminalia* species.

Quantification of Phenolic Compounds in Six *Terminalia* Species by UPLC-QqQ$_{LIT}$-MS/MS

3

3.1 INTRODUCTION

In order to effectively utilize the herbal materials, it is essential to know the distribution of the various phytocomponents in different parts of the plant so that the most suitable part can be used for a particular application. There is no comprehensive study on the quantification and discrimination in terms of the distribution of phenolic compounds in different plant parts of *Terminalia* species. Ultra-performance liquid chromatography-triple quadrupole linear ion trap mass spectrometry (UPLC/QqQ$_{LIT}$-MS) enables rapid detection of the targeted analytes at very low concentration levels due to its selectivity and superior sensitivity with low consumption of solvents and provides fast and robust quantitative data to assess the contents of the various phytoconstituents (Singh et al. 2015; Kong et al. 2014; Hager and Le Blanc 2003). We, therefore, carried out comparative quantitative profiling of phenolic compounds from different plant parts of the *Terminalia* species by liquid chromatography–tandem mass spectrometry with chemometric analysis (Singh et al. 2016b). The aim of this study was the comprehensive identification, quantification and discrimination in terms of distribution of phenolic compounds in different plants parts (bark, fruit, leaf, stem and root) of the six *Terminalia* species, namely, *T. arjuna*, *T. bellerica*, *T. bellirica*, *T. chebula*, *T. elliptica* and *T. paniculata* using UPLC-QqQ$_{LIT}$-MS/MS and 37 phenolic compounds were quantified.

3.2 PLANT MATERIALS AND CHEMICALS

Bark, fruit, leaf, stem and root of the six *Terminalia* species (*T. arjuna*, *T. bellirica*, *T. catappa*, *T. chebula*, *T. elliptica* and *T. paniculata*) were collected from Jawaharlal Nehru Tropical Botanic Garden and Research Institute campus (JNTBGRI; N: 8°45', E: 77°10', Altitude: 70–160 m), Kerala, South India in February 2014. Further details of the samples collected and the chemicals used including the reference compounds were as discussed in Section 2.2.

3.3 EXTRACTION AND SAMPLE PREPARATION

The samples were extracted as given in Section 2.3. The 1 mg/mL stock solution of the extract of each sample was diluted with methanol to working solutions. Using the stock methanol solution (1,000 µg/mL) of reference standards, 1,000 ng/mL mixtures of all the 37 reference standards were prepared in methanol and diluted to appropriate concentrations to yield a series of concentrations, within the range of 0.5–1,000 ng/mL for quantification of phytoconstituents in *Terminalia* species. The calibration curves were constructed by plotting the value of peak areas versus concentrations of each analyte. All stock solutions were stored in the refrigerator at −20°C until use.

3.4 UPLC-QqQ$_{LIT}$-MS CONDITIONS

Quantitative analysis was performed on a 4000 QTRAP™ MS/MS system, hybrid triple quadrupole–linear ion trap mass spectrometer (Applied Biosystems, Concord, ON, Canada), hyphenated with a Waters ACQUITY

UPLC™ system (Waters; Milford, MA, USA) via an electrospray (Turbo V™) interface. The Waters ACQUITY UPLC™ system was equipped with a binary solvent manager, sample manager, column compartment and photodiode array (PDA) detector (Waters, Milford, MA). Chromatographic separation of compounds was achieved with an ACQUITY UPLC BEH™ C18 column (1.7 μm, 2.1 × 100 mm) operated at 35°C. The mobile phase consisted of a 0.1% formic acid aqueous solution (A) and methanol (B), delivered at a flow rate of 0.250 mL/min under a gradient program: 0%–5% (B) initial to 1.0 min, 5%–30% (B) from 1.0 to 5.0 min, 30%–32% (B) from 5.0 min to 7.0 min, 32%–50% (B) from 7.0 to 14.0 min, 50%–90% (B) from 14.0 to 18.0 min, 90%–90% (B) from 18.0 to 19.0 min and back to the initial condition from 19.0 to 22.0 min. The sample injection volume was 2 μL.

For mass spectrometric analysis, the compound-dependent parameters, such as declustering potential (DP), entrance potential (EP), collision energy (CE) and cell exit potential (CXP) were optimized for each compound by direct infusion of 20 ng/mL solutions of each analyte using a Harvard '22' syringe pump (Harvard Apparatus, South Natick, MA, USA) in negative ionization mode. Quadrupole 1 and quadrupole 2 were maintained at unit resolution. Quantitative analysis was performed in multiple reaction monitoring (MRM) mode. The optimized source dependent parameters were as follows: the ion spray voltage (IS) was set to −4,200 V; the turbo spray temperature (TEM) 550°C; nebulizer gas (GS 1) 50 psi; heater gas (GS 2) 50 psi; curtain gas (CUR) 20 psi and collision-activated dissociation gas (CAD) was set at medium and the interface heater was on. High purity nitrogen was used for all processes. AB Sciex Analyst software version 1.5.1 was used to control the LC–MS/MS system and for data acquisition and processing. All the statistical calculations related to quantitative analysis were performed on Graph Pad Prism software version 5.2.5.4. Principal component analysis (PCA) was carried out based on the content (in g/g) of all the quantified bioactive compounds (37) in leaves, stem, root, bark and fruit of the six *Terminalia* species using STATISTICA 7.0 software.

3.5 METHOD DEVELOPMENT

All the 37 reference standards (20 ng/mL) were analyzed by continuous infusion in ESI-MS for the optimization of compound-dependent parameters. All the analytes were detected in negative ion mode as [M-H]⁻ ions within the mass range of m/z 50–2,000 (Figure 3.1). The optimized compound-dependent parameters are given in Table 3.1. Quadrupole 1 and quadrupole 2 were maintained at unit resolution. As listed in Table 3.1, two MRM transitions

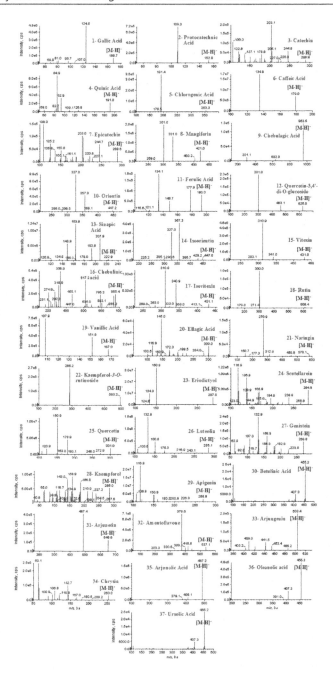

FIGURE 3.1 MS/MS spectrum of the selected analytes.

TABLE 3.1 The optimized compound-dependent MRM parameters and transitions for each analyte in the UPLC-QqQ$_{LIT}$-MS/MS analysis for *Terminalia* species

PEAK NO.	RT (MIN)	ANALYTE	PRECURSOR ION [M-H]⁻	DP (V)	EP (V)	CE (EV)	QUANTIFIER[a]	QUALIFIER[a]
1	2.93	Gallic acid	168.7	−59	−10	−22	168.7→124.8 (−11)	168.7→96.7 (−17)
2	4.27	Protocatechuic acid	152.8	−64	−5	−22	152.8→109.3 (−9)	152.8→91.8 (−2)
3	5.25	Catechin	288.9	−110	−10	−29	288.9→203.1 (−8)	288.9→244.8 (−12)
4	5.58	Quinic acid	191.0	−67	−12	−33	191.0→84.9 (−6)	191.0→92.9 (−15)
5	5.69	Chlorogenic acid	353.3	−60	−10	−30	353.3→191.4 (−10)	353.3→178.5 (−10)
6	6.23	Caffeic acid	179.0	−48	−8	−21	179.0→134.9 (−13)	179.0→107.0 (−8)
7	6.45	Epicatechin	288.6	−120	−9	−29	288.6→203.0 (−6)	288.6→244.7 (−12)
8	6.99	Mangiferin	421.0	−84	−3.2	−32	421.0→301.0 (−14)	421.0→331.0 (−15)
9	7.91	Chebulagic acid	953.5	−176	−10	−10	953.5→953.5 (−16)	953.5→301.1 (−19)
10	9.41	Orientin	447.2	−97	−9	−32	447.2→327.0 (−25)	447.2→357.0 (−8)
11	9.59	Ferulic acid	193.0	−58	−5	−25	193.0→134.0 (−6)	193.0→177.9 (−13)
12	9.63	Quercetin-3,4'-di-O-glucoside	625.0	−97	−6	−43	625.0→301.0 (−7)	625.0→463.1 (−12)
13	9.73	Sinapic acid	222.9	−55	−5	−28	222.9→163.9 (−11)	222.9→207.9 (−8)
14	9.85	Isoorientin	447.0	−97	−8	−32	447.0 →357.3 (−8)	447.0→429.2 (−6)
15	10.83	Vitexin	431.0	−101	−9	−30	431.0→310.9 (−33)	431.0→341.0 (−31)
16	11.60	Chebulinic acid	955.4	−182	−12	−48.9	955.4→336.9 (−16)	955.4→617.2 (−16)
17	11.92	Isovitexin	431.1	−91	−6	−32	431.1→310.9 (−7)	431.1→413.1 (−18)
18	12.36	Rutin	609.4	−197	−10	−50	609.4→300.1 (−25)	609.4→179.3 (−15)
19	12.56	Vanillic acid	167.0	−107	−9.3	−22	167.0→107.9 (−9)	167.0→151.9 −8)

(Continued)

TABLE 3.1 (Continued)   The optimized compound-dependent MRM parameters and transitions for each analyte in the UPLC-QqQ$_{LIT}$-MS/MS analysis for *Terminalia* species

PEAK NO.	RT (MIN)	ANALYTE	PRECURSOR ION [M-H]−	DP (V)	EP (V)	CE (EV)	QUANTIFIER[a]	QUALIFIER[a]
20	12.67	Ellagic acid	300.9	−62	−4	−56	300.9→145.0 (−12)	300.9→199.5 (−11)
21	13.91	Naringin	579.1	−131	−9	−45	579.1→270.9 (−9)	579.1→150.7 (−4)
22	14.00	Kaempferol-3-O-rutinoside	593.2	−105	−5	−45	593.2→285.2 (−6)	593.2→254.7 (−10)
23	14.42	Eriodictyol	287.0	−88	−8.7	−21	287.0→150.8 (−9)	287.0→134.9 (−13)
24	15.09	Scutellarein	284.8	−99	−5.3	−42	284.8→116.9 (−12)	284.8→136.9 (−1)
25	15.64	Quercetin	301.0	−107	−9	−34	301.0→150.9 (−12)	301.0→178.9 (−15)
26	16.19	Luteolin	285.1	−139	−10	−41	285.1→132.9 (−21)	285.1→150.8 (−13)
27	16.40	Genistein	268.8	−105	−8	−43	268.8→132.9 (−7)	268.8→158.9 (−14)
28	16.70	Kaempferol	285.2	−95	−5	−49	285.0→158.9 (−10)	285.0→143.0 (−8)
29	16.80	Apigenin	268.9	−71	−5	−13	268.9 →116.8 (−13)	268.9→106.8 (−8.6)
30	17.15	Betulinic acid	455.2	−140	−9	−7	455.2→455.2 (−15)	455.2→189.0 (−11)
31	17.28	Arjunetin	649.6	−90	−5.2	−7	649.6→487.4 (−13)	649.6→529.1 (−9)
32	17.61	Amentoflavone	537.1	−145	−7	−42	537.1→375.0 (−20)	537.1→416.8 (−10)
33	17.72	Arjungenin	503.4	−123	−9	−7	503.4→503.4 (−15)	503.4→409.0 (−1.8)
34	17.90	Chrysin	253.0	−100	−8	−38	253.0→63.1 (−10)	253.0→142.7 (−8)
35	18.70	Arjunolic acid	487.2	−164	−8	−48	487.2→487.2 (−12)	487.2→409.1 (−11)
36	18.72	Oleanolic acid	455.3	−120	−8	−12	455.3→455.3 (−13)	455.3→407.3 (−16)
37	19.29	Ursolic acid	455.2	−122	−10	−9	455.2→455.2 (−7)	455.2→407.3 (−20)

RT: retention time; DP: declustering potential; EP: entrance potential; CE: collision energy.
[a] Cell exit potential (CXP in V) is given in parentheses.

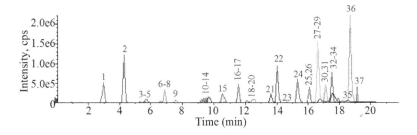

FIGURE 3.2 MRM chromatogram of the 37 reference compounds.

(precursor-to-product ions) were monitored in which the most abundant and stable pair was selected as the quantifier and the other one as the qualifier in quantitative analysis.

During the optimization of chromatographic conditions, methanol displayed stronger resolving power for the isomeric peaks (catechin and epicatechin/ orientin and isoorientin/vitexin and isovitexin/betulinic acid, oleanolic acid and ursolic acid) than acetonitrile.

Separation of isomeric peaks was essential due to the presence of the same qualifier as they were quantified on the basis of their retention time (RT). Hence, 0.1% aqueous formic acid and methanol were chosen as mobile phases for separation of all the compounds within 20 min. The MRM and extracted ion chromatograms obtained for the mixture of 37 standards are given in Figure 3.2.

3.5.1 Analytical Method Validation

The UPLC-ESI–MS/MS method was validated according to the guidelines of the International Conference on Harmonization (ICH, Q2R1) by determining calibration curves, lower limit of detection (LOD), lower limit of quantification (LOQ), precision, solution stability and recovery (ICH Guideline 2005). The linearity, regression and linear ranges of analytes were obtained using the external standard method. The calibration graphs were achieved by the analyte peak area (y) versus concentration (x) and constructed with a weight factor $(1/x^2)$ by least-squares linear regression. Correlation coefficient $(R^2 \geq 0.9972)$ values indicated appropriate correlations between peak area and concentrations of analytes within the test ranges. The LODs and LOQs for each analyte were separately determined at signal-to-noise ratios (S/N) of 3:1 and 10:1, respectively (Table 3.2).

The precision of the method was validated by determination of intraday (n=5) and interday precision (n=9) of the analytes in a single day and three

TABLE 3.2 Linearity, LOD, LOQ, precision, stability and recovery results of 37 investigated components

ANALYTES	REGRESSION EQUATION	R^2	LINEAR RANGE (NG/ML)	LOD (NG/ML)	LOQ (NG/ML)	PRECISION RSD (%) INTERDAY (N=5)	INTERDAY (N=9)	STABILITY RSD (%) (N=5)	RECOVERY MEAN (N=5)	RSD (%)
Gallic Acid	y=5139*x+1155	1.0000	2.25–200	0.74	2.24	0.12	0.21	1.58	104.15	1.21
Protocatechuic Acid	y=11900*x−241.0	0.9999	2.5–200	0.07	0.21	1.23	0.98	2.40	101.58	1.73
Catechin	y=188.1*x−50.96	0.9998	1–500	0.27	0.82	1.74	1.07	2.21	102.32	1.31
Quinic Acid	y=569.7*x+79.30	0.9998	2–250	0.46	1.39	1.39	1.84	2.14	98.01	2.14
Chlorogenic Acid	y=3718*x+856.3	0.9992	0.5–50	0.01	0.03	0.98	1.65	2.06	95.09	0.79
Caffeic Acid	y=6440*x+579	0.9997	20–1000	0.29	0.88	0.61	1.20	2.80	96.95	1.52
Epicatechin	y=160.3*x−89.8	0.9979	5–1000	0.85	2.58	0.51	0.98	1.51	99.42	1.76
Mangiferin	y=1846*x−462.5	0.9999	1.25–200	0.03	0.09	1.32	0.95	0.78	101.66	0.96
Chebulagic Acid	y=651.7*x+201.8	0.9986	0.5–250	0.12	0.36	0.11	1.01	1.56	99.81	2.10
Orientin	y=1711*x+194.2	0.9997	1.5–25	0.37	1.12	2.17	1.99	1.70	100.52	2.46
Ferulic Acid	y=1323*x+226.8	1.0000	2–500	0.57	1.73	1.39	1.50	1.66	98.41	1.08
Quercetin-3,4'-di-O-glucoside	y=1995*x−743.3	0.9999	1–200	0.11	0.33	0.75	1.21	1.89	100.35	2.89
Sinapic Acid	y=1230*x−13.33	0.9998	0.5–250	0.04	0.12	1.88	0.65	2.50	102.65	1.85
Isoorientin	y=3962*x−1137	0.9993	1–125	0.15	0.46	1.69	2.70	1.31	101.63	1.95
Vitexin	y=2484*x−16.78	0.9999	0.5–250	0.02	0.06	0.89	1.50	1.15	105.84	1.88
Chebulinic acid	y=1528*x−1013	1.0000	10–250	2.19	6.45	0.53	3.01	1.32	100.98	1.45
Isovitexin	y=3343*x+937.9	0.9999	1–500	0.13	0.39	1.46	1.16	1.33	97.12	1.65
Rutin	y=268.1*x−122.3	1.0000	1–100	0.51	1.54	2.70	2.65	2.01	98.84	0.92

(Continued)

TABLE 3.2 (Continued) Linearity, LOD, LOQ, precision, stability and recovery results of 37 investigated components

ANALYTES	REGRESSION EQUATION	R^2	LINEAR RANGE (NG/ML)	LOD (NG/ML)	LOQ (NG/ML)	PRECISION RSD (%) INTERDAY (N=5)	INTERDAY (N=9)	STABILITY RSD (%) (N=5)	RECOVERY MEAN (N=5)	RSD (%)
Vanillic Acid	y=226.9*x+139.2	0.9972	1.25–250	0.22	0.67	2.12	2.14	1.25	97.27	1.12
Ellagic Acid	y=107.7*x+30.16	1.0000	1.25–100	0.14	0.42	1.62	1.56	2.14	100.63	1.01
Naringin	y=2350*x+91.29	1.0000	1–200	0.13	0.39	2.01	2.10	1.80	100.08	0.62
Kaempferol-3-O-rutinoside	y=3147*x−824.4	0.9984	1.25–50	0.16	0.48	0.57	2.50	2.10	98.89	3.01
Eriodictyol	y=12264*x−5551	0.9995	5–12.5	0.24	0.73	0.47	1.98	1.85	99.01	1.85
Scutellarein	y=609.7*x−388.7	0.9995	10–200	1.10	3.33	0.43	0.89	1.25	100.15	2.14
Quercetin	y=2665*x+754.0	1.0000	1–250	0.33	1.00	0.90	0.99	2.14	96.12	1.20
Luteolin	y=1106*x+770.0	1.0000	5–200	0.36	1.09	1.62	1.81	1.97	104.20	0.98
Genistein	y=4986*x−869.8	0.9988	1.5–20	0.58	1.76	1.21	1.51	1.85	102.40	0.32
Kaempferol	y=419.4*x−49.79	0.9999	1–100	0.39	1.18	0.63	0.87	1.87	98.71	1.26
Apigenin	y=15836*x−340.0	0.9999	1.25–500	0.07	0.21	0.53	0.95	0.15	95.62	1.36
Betulinic Acid	y=49558*x+5236	1.0000	1.5–100	0.35	1.06	1.25	1.10	1.63	101.86	2.96
Arjunetin	y=373.0*x−164.8	0.9999	5–250	1.46	4.42	1.40	1.49	1.90	102.75	2.60
Amentoflavone	y=4944*x+525.1	0.9987	1.5–25	0.35	1.06	1.68	1.85	1.04	105.69	1.20
Arjungenin	y=18051*x+2237	0.9998	1.5–100	0.41	1.24	1.52	3.21	1.99	104.25	0.48
Chrysin	y=3482*x+1916	0.9999	5–1000	0.82	2.48	0.33	1.21	0.95	98.05	0.89
Arjunolic Acid	y=20363*x+3487	1.0000	2.5–125	0.57	1.73	0.14	1.30	1.48	99.95	1.45
Oleanolic acid	y=21366*x+6908	1.0000	5–200	1.07	3.24	0.98	1.54	1.26	98.31	2.65
Ursolic Acid	y=6799*x+2803	0.9999	1–250	0.36	1.09	1.47	2.92	3.14	101.16	2.01

y: peak area; **x**: concentration of compound (ng/ml); **LOD**: limit of detection: S/N=3.3; **LOQ**: limit of quantification: S/N=10.

consecutive days, respectively. The intraday and interday precisions calculated as relative standard deviation (RSD) were in the range of 0.11%–2.70% and 0.21%–3.21%, respectively. Stability was tested at room temperature (22°C–24°C), and samples were analyzed within 48 h. The stability RSD values of the analytes were in the range of 0.15%–3.14%. A recovery test was used to evaluate the accuracy of the method. The recoveries were determined by spiking accurately known amounts of all the analytes (high, middle and low) into a sample followed by extraction and analysis using the described method. The recovery was calculated by the formula: recovery = $(a − b)/c × 100$%, where 'a' is the detected amount, 'b' is the original amount and 'c' is the spiked amount. The recovery of the method was in the range of 95.09%–105.84%, with RSD less than 3.01%, indicating that the established method was accurate for the determination of the selected compounds in *Terminalia* species.

3.6 QUANTITATIVE ANALYSIS FOR PHENOLICS

The content of each analyte in different plant parts of *Terminalia* species was calculated from the corresponding calibration curve and summarized in Table 3.3, which indicates significant variations among the contents (Figure 3.3). The total content of the 37 constituents in the extracts of bark, fruit, leaf, stem, and root taken together was the highest in *T. chebula* and the lowest in *T. paniculata* in the following order: *T. chebula* (11.14% w/w)>*T. elliptica* (6.28% w/w)>*T. arjuna* (4.61% w/w)>*T. bellirica* (4.4% w/w)>*T. catappa* (2.14% w/w)>*T. paniculata* (1.37% w/w). The part-wise distribution of the contents showed that the highest total content of 37 analytes was in *T. chebula* fruit (29.41% w/w) followed by *T. arjuna* root (11.70% w/w), *T. elliptica* leaf (10.99% w/w) and *T. chebula* leaf (10.54% w/w) (Figure 3.3).

Among the fruits, *T. chebula* (29.41% w/w) showed the highest total content, whereas *T. bellirica* (10.18% w/w) and *T. elliptica* (10.32% w/w) showed a significant total content of 37 compounds. In *T. catappa* and *T. paniculata*, it was the leaf that showed the highest total content of 5.39% w/w and 3.75% w/w, respectively. *T. elliptica* showed almost similar contents in the leaf (10.99% w/w) and the fruit (10.32% w/w). It is the root that showed the highest content (11.7% w/w) in *T. arjuna*. Its bark (3.69% w/w) and fruit (3.6% w/w) had similar contents, whereas the leaf (2.44% w/w) and stem (1.61% w/w) had much lesser contents in *T. arjuna*. In *T. bellirica*, the leaf and stem showed similar total contents of 4.91% w/w and 4.86% w/w, respectively, whereas the roots (1.65% w/w) showed much less.

TABLE 3.3 Contents (in μg/g) of 37 bioactive compounds in *Terminalia* species

ANALYTES (μg/g)	TAJ L	TAJ S	TAJ R	TAJ B	TAJ F	TBL L	TBL S	TBL R	TBL B	TBL F	TCB L	TCB S	TCB R	TCB B	TCB F
Gallic Acid	740	255	394.5	492.5	175.5	13780	245	605	77.5	26750	65.5	324	986	1340	13800
Protocatechuic Acid	29	10.45	12.9	442.5	31.1	39.3	44.3	36.15	34.1	104.5	31.75	21.15	26.1	59.5	505
Catechin	580	1925	1435	640	485	613.2	1310	393	49	35.9	2680	358.5	89.1	2130	49
Quinic Acid	101	82	161	36.65	1970	798	446	535	467.5	6450	3130	4610	2987	4895	16900
Chlorogenic Acid	41	17	bdl	bdl	bdl	bdl	bdl	bdl	bdl	bdl	0.78	0.39	0.36	bdl	11.9
Caffeic Acid	0.07	0.12	bdl	bdl	bdl	0.17	0.39	bdl	bdl	bdl	0.051	0.14	0.101	bdl	bdl
Epicatechin	449.5	266	358	470	220	514	615	1140	235.5	225	276	220	199.1	296.5	245.5
Mangiferin	nd	19.45	nd	16.8	15.05	nd	nd	17.7	16.8	15.9	15.9	15.05	14.01	15.9	17.7
Chebulagic Acid	96	32.9	nd	214.5	144	29012.1	33350	242	215.5	36650	6350	63	51	2.645	163500
Orientin	52.5	nd	bdl	bdl	5.9	nd	30.8	nd	0.625	bdl	660	0.1485	0.014	3.025	18.35
Ferulic acid	75.5	2.9	bdl	15.3	9.1	13.9	24.6	278	101.5	34.5	14.05	12.8	10.52	6.6	20.25
Quercetin-3,4'-di-O-glucoside	22.2	18.1	nd	19.75	18.9	nd	22.2	nd	23.85	19.75	18.1	24.65	11.4	18.9	21.35
Sinapic Acid	3.48	11.5	1.47	3.475	2.14	bdl	9.5	3.475	nd	12.15	nd	3.475	2.14	5.5	2.14
Isoorientin	66	17.5	17.5	19.15	22.85	34.8	41.9	16.85	16.05	23.3	975	16.65	9.25	15.2	35.9
Vitexin	665	5.75	1.8	8.75	20.3	198	4.44	4.44	2.46	nd	455.5	1.8	1.09	2.46	26.55
Chebulinic Acid	133.5	216	312	190	248.5	952	4950	446	101.5	8150	510	381	229	338	59500
Isovitexin	590	bdl	bdl	69	bdl	bdl	bdl	bdl	bdl	bdl	1865	bdl	3.001	bdl	4.005

(Continued)

TABLE 3.3 (*Continued*)　Contents (in µg/g) of 37 bioactive compounds in *Terminalia* species

ANALYTES (µg/g)	TAJ L	TAJ S	TAJ R	TAJ B	TAJ F	TBL L	TBL S	TBL R	TBL B	TBL F	TCB L	TCB S	TCB R	TCB B	TCB F
Rutin	144.5	34.45	nd	40.55	157	101	92.5	nd	nd	nd	224.5	46.7	12	65	59
Vanillic Acid	80	7.7	11.3	0.45	14.9	34	47.45	156	14.9	127	bdl	nd	nd	36.6	bdl
Ellagic Acid	2095	1285	2820	2150	4725	1982	5550	10950	565	22900	4685	16650	15264	16600	30100
Naringin	0.17	0.012	nd	nd	0.007	bdl	bdl	nd	0.78	0.14	0.09	bdl	0.07	bdl	0.011
Kaempferol-3-O-rutinoside	27.85	12.45	12.45	12.45	14.55	18.01	20.8	12.45	12.45	14.55	32.85	12.45	10.4	13.5	14.05
Eriodictyol	28.4	54	27.1	23	28.4	79.2	125	104.5	29.75	24.55	23.5	29.95	2.14	48.4	24.15
Scutellarein	nd	nd	47.2	33.75	nd	32.18	36.45	33.75	36.45	52.5	49.9	44.5	21.5	41.85	44.5
Quercetin	80.5	bdl	2.45	8	bdl	bdl	bdl	18.45	bdl	bdl	bdl	4.91	2.6	bdl	4.91
Luteolin	88	bdl	bdl	bdl	bdl	4.9	5.1	bdl	bdl	bdl	10	bdl	0.012	bdl	bdl
Genistein	10.9	9.6	9.9	9.1	11.4	11.12	10.25	9.25	10.05	9.6	9.6	10.4	8.7	11.55	9.25
Kaempferol	68.5	17.65	17.65	17.65	17.65	41.21	48.95	nd	19.65	29.4	29.4	37.2	15	25.5	17.65
Apigenin	8.85	7.4	6.5	5.7	8.85	6.92	8.95	4.3	6.35	4.405	6.55	7	4.98	5.55	6.55
Betulinic Acid	229.5	2385	30700	965	165.5	91.8	105	3.045	2.125	bdl	4300	1100	952	87	2.685
Arjunetin	4945	2440	23350	12450	bdl	97.1	169.5	42	237.5	116	5200	510	135.8	910	476
Amentoflavone	bdl	bdl	bdl	bdl	bdl	0.1422	0.1755	bdl	bdl	bdl	0.505	bdl	0.014	bdl	0.1755
Arjungenin	2385	2250	29700	14700	6200	31	47.2	61.5	399	46.7	10600	27500	17560	22050	7150
Chrysin	0.14	1.36	1.78	0.98	0.21	0.47	0.51	0.1	bdl	bdl	0.36	0.14	0.19	0.21	1.1
Arjunolic Acid	9350	4415	25750	3720	4450	119.1	431	1040	1470	bdl	41800	5850	9850	1580	1490
Oleanolic acid	515	233	1035	81	349.5	298	370	238	3.945	bdl	6450	121.5	98.1	15.55	bdl
Ursolic Acid	730	116.5	770	67	457	199.1	408.5	87	bdl	bdl	14900	115	14.25	bdl	bdl

(*Continued*)

TABLE 3.3 (Continued) Contents (in μg/g) of 37 bioactive compounds in *Terminalia* species

ANALYTES (μg/g)	TAJ L	TAJ S	TAJ R	TAJ B	TAJ F	TBL L	TBL S	TBL R	TBL B	TBL F	TCB L	TCB S	TCB R	TCB B	TCB F
	TCT L	*TCT S*	*TCT R*	*TCT B*	*TCT F*	*TPN L*	*TPN S*	*TPN R*	*TPN B*	*TPN F*	*TEP L*	*TEP S*	*TEP R*	*TEP B*	*TEP F*
Total %	2.44	1.61	11.70	3.69	3.60	4.91	4.86	1.65	0.41	10.18	10.54	5.81	4.86	5.06	29.41
Gallic Acid	174.5	510	3130	240.5	135.5	1110	284.5	238.5	89.5	314.5	1712.6	236.2	194.5	30.15	1275
Protocatechuic Acid	78	20.9	69	20.1	90	44.85	30.65	35.75	13.5	71.4	61.9	35.6	53.5	23.35	315
Catechin	101.5	965	197	214.5	49	179.5	493.5	275.5	1680	178.8	2987.1	1041	4025	12550	565
Quinic Acid	185.5	61.5	55.5	29.85	520	166	156	625	106.5	98.2	251	142	54.5	52	33.95
Chlorogenic Acid	1.17	bdl	bdl	nd	bdl	0.013	bdl	bdl	bdl	bdl	bdl	bdl	bdl	0.825	bdl
Caffeic Acid	0.014	bdl	bdl	bdl	bdl	0.123	0.036	0.147	1.36	0.25	1.25	bdl	bdl	bdl	bdl
Epicatechin	220	281.5	225	nd	235.5	240.5	235.5	220	276	142	165.8	98	266	550	235.5
Mangiferin	15.05	nd	20.35	nd	16.8	nd	22.15	nd	23	17.8	12	0.98	19.45	nd	nd
Chebulagic Acid	20.3	35.4	55.5	27.85	451.5	68	116	142.5	43	28.7	78.9	47	bdl	184	81
Orientin	7.35	nd	7.8	bdl	7.8	bdl	8.55	bdl	1.105	0.39	1.56	1.47	3.02	bdl	24.1
Ferulic acid	17.75	19	54.5	bdl	bdl	127	48.75	7.85	50.5	11	2.5	0.25	7.85	5.4	64
Quercetin-3,4'-di-O-glucoside	18.1	18.9	32.05	19.75	19.75	23.85	20.55	18.9	18.9	17.8	1.8	1.7	26.3	24.65	18.9
Sinapic Acid	3.475	4.81	10.15	11.5	8.8	21.5	6.15	6.15	4.81	2.5	14	11.8	16.85	3.475	15.5
Isoorientin	41.3	15.4	17.05	16.25	23.1	19.55	24.5	17.5	15.8	11.7	14.01	20.08	17.25	15.4	30.25
Vitexin	133.5	1.135	4.44	nd	1.8	4.77	93	nd	2.46	1.4	0.78	0.14	0.805	1.135	117
Chebulinic Acid	184	459	590	530	111.5	234	600	755	349	214	111	124	210	188.5	95

(Continued)

TABLE 3.3 (Continued)   Contents (in μg/g) of 37 bioactive compounds in *Terminalia* species

ANALYTES (μg/g)	TAJ L	TAJ S	TAJ R	TAJ B	TAJ F	TBL L	TBL S	TBL R	TBL B	TBL F	TCB L	TCB S	TCB R	TCB B	TCB F
Isovitexin	580	bdl	bdl	bdl	bdl	nd	49.65	bdl	bdl	1.78	0.089	0.14	bdl	bdl	91.5
Rutin	138.5	62	nd	34.45	108	40.55	40.55	34.45	73.5	14.5	0.17	1.4	nd	nd	77.5
Vanillic Acid	nd	76.5	25.75	40.2	22.15	4.06	18.5	36.6	47.45	25	14	58.4	62	4.06	112.5
Ellagic Acid	5150	1130	2265	705	2470	4480	1730	2215	5450	3951	890	485	1970	795	6100
Naringin	bdl	nd	nd	nd	bdl	bdl	bdl	0.036	bdl	0.14	0.051	0.14	0.012	0.07	bdl
Kaempferol-3-O-rutinoside	25.55	12.45	nd	nd	16.65	14.05	nd	16.1	nd	15	10.6	9.85	13.5	nd	nd
Eriodictyol	23.3	32	46.55	nd	29.4	32.95	27.6	47.75	23.7	12	14.5	10	33.35	23.5	29.2
Scutellarein	36.45	nd	nd	nd	49.9	33.75	nd	41.85	33.75	0.98	14	12.4	41.85	36.45	41.85
Quercetin	22.75	17.85	bdl	nd	12.3	74	bdl	bdl	bdl	nd	bdl	bdl	bdl	bdl	14.75
Luteolin	16	bdl	bdl	nd	bdl	3.16	3.63	bdl	bdl	1.2	2.1	1.1	bdl	bdl	bdl
Genistein	9.9	10.9	12.4	9.4	10.25	10.25	10.25	9.9	10.25	8.9	8.9	7.4	10.9	9.4	10.55
Kaempferol	21.6	19.6	60.5	51	88	43.1	29.4	25.5	nd	0.014	10.58	12.4	17.65	nd	11.8
Apigenin	6.35	6.5	4.3	5.7	7.1	6.75	15.85	5.5	4.92	1.45	3.9	2.8	7.2	4.2	10.45
Betulinic Acid	1180	424.5	3360	7550	42.7	272.5	1535	710	545	79	987.1	254.1	1165	665	635
Arjunetin	3560	2310	3580	1525	46.55	2275	580	775	134.5	254.8	21578	9801	10200	8500	7700
Amentoflavone	bdl	bdl	bdl	bdl	bdl	bdl	bdl	bdl	bdl	bdl	0.014	bdl	bdl	bdl	bdl
Arjungenin	995	1655	317.5	221.5	760	419.5	90	24.6	7.9	3.6	1985	940	2055	2740	54500
Chrysin	1.23	1.42	bdl	1.41	1.71	2.7	nd	nd	bdl	0.14	2	0.25	2.61	1.87	2.02
Arjunolic Acid	39050	5650	3130	1810	1930	25600	2190	1405	bdl	98	78510	1459.1	34900	1290	25850
Oleanolic acid	840	199.5	76	250.5	313.5	665	108.5	20.15	14	11.2	154.6	1235	222	bdl	2250
Ursolic Acid	1020	100	30.85	428	308	1245	6.35	bdl	0.11	0.39	254	1458	124.5	2.655	2865
Total %	5.39	1.41	1.74	1.37	0.79	3.75	0.86	0.77	0.90	0.56	10.99	1.75	5.57	2.77	10.32

TAJ – *T. arjuna*; TBL – *T. bellirica*; TCB – *T. chebula*; TCT – *T. catappa*; TPN – *T. paniculata*; TEP – *T. elliptica*; L – leaf; R – root; S – stem; B – bark and

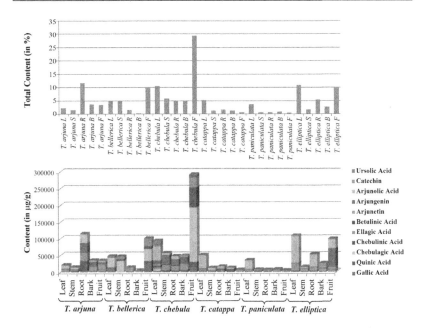

FIGURE 3.3   (a) Total content of 37 analytes (%) detected in different parts of *Terminalia* species. (b) Graphical representation of major bioactive compounds detected in different parts of *Terminalia* species.

Among the different plant parts of the six *Terminalia* species, the least total content (0.41% w/w) was seen in *T. bellirica* bark.

The leaves of *T. chebula* showed a total content of 10.54% w/w, whereas its stem, root and bark had 5.81% w/w, 4.86% w/w and 5.06% w/w, respectively. The stem, root and bark of *T. catappa* showed almost similar contents of 1.41% w/w, 1.74% w/w and 1.37% w/w, respectively. A much lesser content (0.79% w/w) was observed in the fruits. Among the fruits, the total content (0.56% w/w) was the lowest in *T. paniculata*, whereas its stem, root and bark had similar total contents of 0.86% w/w, 0. 77% w/w and 0.9% w/w, respectively. The stem, root and bark of *T. elliptica* showed total contents of 1.75% w/w, 5.57% w/w and 2.77% w/w, respectively. Furthermore, gallic acid (**1**), catechin (**3**), quinic acid (**4**), chebulagic acid (**9**), chebulinic acid (**16**), ellagic acid (**20**), betulinic acid (**30**), arjunetin (**31**), arjungenin (**33**), arjunolic acid (**35**), oleanolic acid (**36**) and ursolic acid (**37**) were detected as the major bioactive constituents (Figure 3.3, Table 3.3). As shown in Figure 3.3, gallic acid was estimated maximum in *T. bellirica* fruit (26,750 µg/g), whereas catechin was found maximum in *T. elliptica* bark (12,550 µg/g). Among the six *Terminalia* species, the highest quantity (163,500 µg/g) estimated for a constituent was for chebulagic acid in *T. chebula* fruit. The maximum estimated quantities of quinic acid, chebulinic acid and ellagic acid were 16,900, 59,500 and 30,100 µg/g, respectively, in *T. chebula* fruit. Betulinic acid, a well-known anticancer compound, and arjunetin were detected highest (30,700 and 23,350 µg/g) in *T. arjuna* root. On the other hand, arjungenin and arjunolic acid, previously quantified in *T. arjuna* stem-bark (Kalola and Rajani 2006; Singh et al. 2002) were found maximum in *T. elliptica* fruit (54,500 µg/g) and leaf (78,510 µg/g), respectively.

Likewise, ursolic acid was found maximum in *T. chebula* leaf (14,900 µg/g). Chebulagic acid was quantified as the most prominent compound among *Terminalia* species.

3.7 CHEMOMETRIC ANALYSIS

The UPLC-ESI–MS/MS data of the plant parts of *Terminalia* species were subjected to PCA to evaluate the variation among the six *Terminalia* species and simplify the data organization. It is an unsupervised clustering method without any information of the data set and retains maximum variance of multi-dimensional data while reducing its dimensionality. The data are generally presented as a two-dimensional plot (score plot) where the coordinate axis represents the directions of the two largest variations.

In order to get a better understanding of different *Terminalia* species, the multivariate study has been done on each plant part separately. The matrix of the peak was reduced to 18 on the basis of their contribution to the PCA scoring. On the basis of eigen values, gallic acid, protocatechuic acid, quinic acid, chebulagic acid, orientin, sinapic acid, isoorientin, vitexin, chebulinic acid, isovitexin, rutin, ellagic acid, kaempferol-3-O-rutinoside, luteolin, amentoflavone, arjunolic acid, oleanolic acid and ursolic acid were observed with covariance 50.77% and 21.40%, which amounts to 72.17% variation for leaf data; covariance 38.54% and 23.87% amounting to 62.41% variation for stem data; covariance 37.25% and 25.35% amounting to 62.60% variation for root data; covariance 42.49% and 36.77% amounting to 79.26% variation for fruit data; in factor 1 versus factor 2 (i.e., PC1 vs PC2) from the six *Terminalia* species. In the bark of all the *Terminalia* species isovitexin, luteolin and amentoflavone were absent. Therefore, 15 peaks out of 18 were observed with covariance 36.65% and 31.69% and together contributed 68.34% variation in PC1 versus PC2 for bark data. The score plots (PC1 vs PC2) for the leaf, stem, root, bark and fruit are given in Figure 3.4. In the PCA plot of leaf, *T. catappa*, *T. paniculata*, *T. elliptica* and *T. bellirica* are located to the right of the vertical axis representing PC1 (positive PC1 values), whereas *T. arjuna* and *T. chebula* are positioned to the left side. *T. catappa*, *T. paniculata*, and *T. elliptica* leaves are positioned closely showing similarity in their pattern of constituents (Figure 3.4a). In the stem part, the PC1 of *T. chebula*, *T. catappa*, *T. paniculata* and *T. arjuna* is associated with positive values, whereas *T. elliptica* is positioned close to the vertical axis but showing negative PC1. *T. bellirica* stem is placed to the negative PC1 and positive PC2 (Figure 3.4b). Among the six species, *T. paniculata* and *T. arjuna* stems showed similarity in their pattern of constituents in negative PC2, whereas *T. catappa* and *T. chebula* stems showed similarity in positive PC2. *T. arjuna*, *T. catappa*, *T. elliptica*, *T. paniculata* and *T. bellirica* roots were located to the left of the vertical axis representing

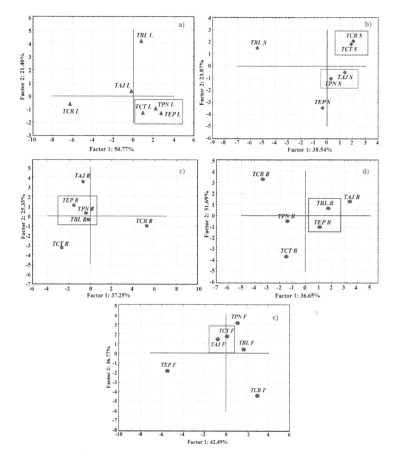

FIGURE 3.4   Score plot of factor 1 versus factor 2 (PC1 vs PC2) scores: (a) leaf, (b) stem, (c) root, (d) bark and (e) fruit parts of the six *Terminalia* species.

PC1, whereas *T. chebula* root was positioned to the right of PC1 (Figure 3.4c). Out of the six species, the roots of *T. elliptica, T. paniculata* and *T. bellirica* were placed very close to each other indicating similarity in their constituents. In the case of bark, *T. bellirica, T. arjuna* and *T. elliptica* were located at the right side of the vertical axis (positive PC1 values), while those of *T. chebula, T. paniculata* and *T. catappa* were positioned to the negative side of PC1. As the barks of *T. bellirica* and *T. elliptica* were close with respect to their PC1 values, they must be having similarities in their pattern of constituents. The PC scores of barks of these species were found scattered in the plot (Figure 3.4d) indicating their distinctive character in the quantity of compounds. In the case of fruit, *T. paniculata, T. bellirica* and *T. chebula* showed positive PC1, while *T. arjuna* and *T. elliptica* showed negative PC1; *T. catappa* fruit was positioned on the vertical axis toward the positive PC1 value; *T. catappa* and *T. arjuna* were positioned close to each other (Figure 3.4e). The scores of replicate measurements were more or less superimposable as the triplicate measurements from the same sample were found to be highly reproducible. The score plots for the leaf, stem, root, bark and fruit showed clustering and differentiation of the data according to the species. It is, therefore, clear that UPLC-ESI–MS/MS followed by PCA is an appropriate method for the identification of different species.

3.8 CONCLUSION

Significant differences were observed in the contents of the 37 phytoconstituents in the six species of *Terminalia* investigated. The contents varied in the plant parts leaf, stem, root, bark and fruit. The total content of all the 37 compounds was the highest in *T. chebula* fruit. All the parts showed significant contents in *T. chebula*, whereas the root, bark and fruit of *T. arjuna* showed more content than its leaf and stem. The root and bark were less important in *T. bellirica*, whereas its fruit had the highest content equal to the combined contents of its leaf and stem. The content of the leaves of *T. catappa* was almost equal to the contents of its stem, root, bark and fruit together. This is also true for *T. paniculata*. On the other hand, *T. elliptica* showed significant contents, especially in the leaves, root, bark and fruit with the leaf and fruit having almost similar contents. PCA analysis could discriminate among the different plant parts of the six *Terminalia* species. The quantitative results also indicated the importance of the various plant parts in terms of the detected bioactive phytoconstituents. Comprehensive estimation of selected bioactive compounds could make a contribution to the quality control of *Terminalia* species.

Simultaneous Estimation of Phytoconstituents in *T. chebula* Fruit and Its Marketed Polyherbal Formulations

4

4.1 INTRODUCTION

One of the fundamental requirements of the herbal industry is the identification and quality evaluation of crude herbs and products. These are more important when the fact is taken into account that the plant material to be examined usually has a complex and inconsistent composition based on various factors such as geographical location and climatic and soil conditions. The genus *Terminalia* is a rich source of tannins (Walia and Arora 2013). Its content can vary with seasonal and environmental factors. The tannin content was shown to depend on the geographical location in *T. chebula* (Kumar 2006). Identification and evaluation of raw materials have become the fundamental need of the herbal industry to ensure that raw materials are authentic and of prescribed quality (Mukherjee 2002; Mukherjee et al. 2008). We, therefore, decided to investigate the effect of geographic location on the distribution of phytoconstituents in *T. chebula* fruits collected from three different locations in three consecutive years by simultaneously determining the quantitative contents of 17 bioactive phytoconstituents using UPLC-QqQ$_{LIT}$-MS. The phytochemical contents of several polyherbal formulations collected from the local market were also determined.

4.2 PLANT MATERIALS AND CHEMICALS

The fruits of *T. chebula* were collected from Lucknow (Uttar Pradesh), Jabalpur (Madhya Pradesh) and Kolkata (West Bengal) in India during the years 2010, 2011 and 2012. Voucher specimens of *T. chebula* have been deposited in the Botany Department of CSIR-Central Drug Research Institute, Lucknow. The specimen voucher number, sample code, collection year and location are provided in Table 4.1. Marketed polyherbal formulations HF1-HF9 were procured from the local market. Further details of the chemicals and reference samples used were as discussed in Section 2.2.

TABLE 4.1   Voucher specimen number, sample code, collection year and location of *Terminalia chebula* fruit

S. NO.	VOUCHER SPECIMEN NUMBER	LOCATION IN INDIA	SAMPLE CODE	COLLECTION YEAR
1	24528L	Lucknow (Uttar Pradesh), 26°52'24.92" N 80°52'26.74" E, Elev 392 ft.	TC1 TC2 TC3	2010 2011 2012
2	24534J	Jabalpur (Madhya Pradesh), 23°11'33.71" N 79°55'52.63" E, Elev 1305 ft.	TC4 TC5	2010 2011
3	24525K	Kolkata (West Bengal), 22° 26'21.14" N 88°24'09.21" E, Elev 26 ft.	TC6 TC7 TC8	2010 2011 2012

4.3 EXTRACTION AND SAMPLE PREPARATION

The extraction and sample preparation of the fruits of *T. chebula* from the different locations were as described in Section 2.3. The samples of the poly-herbal formulations HF1-HF9 were extracted and prepared as follows. Each tablet of polyherbal formulations of *T. chebula* fruit was finely powdered after removing the coating. The sample powder (250 mg) was weighed and soni-cated using an ultrasonicator (Bandelin SONOREX, Berlin) using 25 mL of 100% methanol at room temperature for 1 h and filtered through a 0.22 μm syringe filter (Millex-GV, PVDF, Merck Millipore, Darmstadt, Germany) to obtain 10,000 μg/mL. The filtrates were diluted with methanol to final work-ing solutions and analyzed directly by UPLC-ESI-MS/MS.

The methanol stock solution containing the mixture of 17 analytes (gallic acid, protocatechuic acid, (+)-catechin, quinic acid, (+)-epicatechin, chebu-lagic acid, chebulinic acid, rutin, ellagic acid, eriodictyol, scutellarein, kaemp-ferol, apigenin, betulinic acid, arjunetin, arjungenin and arjunolic acid) was prepared and diluted in appropriate concentration to yield a series of concen-trations from 0.5 to 1,000 ng/mL for simultaneous determination of multiple bioactive phytoconstituents in *T. chebula* fruit from different locations and its marketed polyherbal formulations. The calibration curves were constructed by plotting the value of peak areas versus concentrations of each analyte. All stock solutions were stored in the refrigerator at −20°C until use.

4.4 QUANTITATIVE ANALYSIS

4.4.1 Ultra-performance Liquid Chromatography/Triple Quadrupole Linear Ion Trap Mass Spectrometry (UPLC-QqQ$_{LIT}$-MS) Conditions

The UPLC and QqQ$_{LIT}$-MS conditions were as described in Section 3.3. Chromatographic separation of compounds was obtained with an ACQUITY UPLC BEH™ C18 column (1.7 μm, 2.1×50 mm) operated at 45°C. The mobile phase consisted of a 0.1% formic acid aqueous solution (A) and methanol (B), delivered at a flow rate of 0.350 mL/min under a gradient program: 0%–8% (B) initial to 0.6 min, 8%–20% (B) from 0.6 to 3.0 min, 20%–22% (B) from 3.0 to 5.0 min, 22%–24% (B) from 5.0 to 6.0 min, 24%–34% (B) from 6.0 to 7.0 min, 34%–45% (B) from 7.0 to 8.0 min, 45%–55% (B) from 8.0 to 12.0 min, 55%–70% (B) from 12.0 to 13.0 min, 70%–90% (B) from 13.0 to 13.5 min, 90%–90% (B) from 13.5 to 14.0 min and back to the initial condition. The sample injection volume was 1 μL.

4.4.2 Method Development

A total of 17 reference standards were analyzed by ESI-MS/MS analysis for quantification in different samples of *T. chebula* fruit. All the 17 reference standards were detected in negative ion mode as [M-H]⁻ ions within the mass range of m/z 50–2000 as discussed in Section 3.4.

4.4.3 Analytical Method Validation

The proposed UPLC-ESI-MS/MS method for quantitative analysis was validated as discussed in Section 3.4.3 according to the ICH, Q2R1 guidelines by determining calibration curves, lower limit of detection (LOD), lower limit of quantification (LOQ), specificity, precision, solution stability and recovery (ICH Guideline, 2005). The linearity, regression and linear ranges of the 17 analytes were obtained using the external standard method. The linearity of calibration was achieved by the analyte peak area (y) versus concentration (x) and constructed with a weight factor ($1/x^2$) by least-squares linear regression. All of the correlation coefficient ($R^2 \geq 0.9990$) values indicated appropriate correlations between peak area and concentrations of analytes within the test ranges.

4.4.4 Estimation of the 17 Phenolics and Terpenoids in Extracts of *T. chebula* Fruits and Polyherbal Formulations

The contents of each analyte in *T. chebula* fruit samples and its marketed polyherbal formulations were calculated from the corresponding calibration curve and are summarized in Table 4.2. Graphical representation of this observation given in Figure 4.1 shows that the concentration of the 17 bioactive compounds varied significantly in the ethanolic extracts of *T. chebula* fruit

TABLE 4.2 Contents (in μg/g) of the 17 bioactive compounds in *Terminalia chebula* and its polyherbal formulations

ANALYTES (μg/g)	TERMINALIA CHEBULA FRUIT								POLYHERBAL FORMULATIONS							
	TC1	TC2	TC3	TC4	TC5	TC6	TC7	TC8	HF1	HF2	HF3	HF4	HF5	HF6	HF7	HF8
Gallic acid	24000	16200	32200	13400	13800	9720	13700	15500	621	394	534	143	320	433	585	102
Protocatechuic acid	624	2370	517	1900	505	699	2150	786	23	14.3	57.9	36.3	5.73	8.2	35.3	8.61
Catechin	1150	nd	3070	806	49	1240	1070	nd	1.85	2.35	2.14	1.34	0.55	0.13	2.07	4.86
Quinic acid	21500	18200	18200	18000	16900	22200	17800	13600	428	443	1000	142	212	149	602	279
Epicatechin	4710	4890	4600	4710	245.5	5110	5520	5110	1.02	1.63	1.04	0.97	0.53	0.53	2.06	0.83
Chebulagic acid	182000	145000	48700	167000	163500	374000	271000	201000	1.27	207	1.19	0.56	nd	0.13	3990	1.41
Chebulinic acid	43500	49600	12700	85700	59500	24000	43900	36100	11.9	195	3.26	4.38	3.33	nd	1420	nd
Rutin	1180	nd	750	1060	59	nd	811	934	nd	nd	nd	nd	nd	0.08	0.16	nd
Ellagic acid	76400	90500	59800	42000	30100	72500	107000	83000	352	224	4.25	312	110	4.13	653	5.91
Eriodictyol	470	511	563	492	24.15	531	491	nd	nd	nd	0.11	nd	0.05	0.06	nd	0.17
Scutellarein	890	917	890	nd	44.5	nd	1050	890	0.09	nd	0.22	0.14	0.40	0.08	nd	0.11
Kaempferol	549	nd	822	432	17.65	510	432	432	1.4	1.72	0.20	32.4	0.16	0.13	0.36	0.23
Apigenin	695	645	687	672	6.55	600	666	642	0.08	0.12	0.714	0.51	1.99	0.08	0.73	0.34
Betulinic acid	129	bdl	bdl	nd	2.69	nd	bdl	1210	11.3	70.5	0.29	65.8	1.73	0.36	2.98	bdl
Arjunetin	4100	4220	2420	2740	476	9820	9100	3570	484	112	41.9	17.3	114	27.7	39.6	19.1
Arjungenin	16000	8290	19000	15900	7150	21400	36400	25700	36.9	99.1	9.07	87.7	5.03	2.37	175	46.2
Arjunolic acid	4390	2850	4110	bdl	1490	3570	4470	3560	13.9	13.2	5.56	21.7	1.73	0.25	18.2	14.1
Total	382287	344193	209029	354812	293870	545900	515560	392034	1987.71	1777.9	1661.86	866.1	777.22	626.22	7526.45	482.87

nd – not detected; bdl – below detection limit.

collected from different geographical locations and also in the marketed poly-herbal formulations.

In *T. chebula* fruit extracts of the samples collected from different locations, the total content of the 17 bioactive compounds was found to be the highest in the sample from Kolkata (TC6, 545,900 µg/g) and lowest in the sample from Lucknow (TC3, 209,029 µg/g) region. Furthermore, chebulagic acid was found to be the most abundant compound (≥145,000 µg/g) in the *T. chebula* fruit samples except for TC3 (48,700 µg/g) wherein ellagic acid (59,800 µg/g) was found to be the most abundant bioactive compound. Chebulagic acid was detected maximum in TC6 (374,000 µg/g) and minimum in TC3. Chebulinic acid was in the range of 12,700–85,700 µg/g. Gallic acid, quinic acid, chebulinic acid, ellagic acid and arjungenin were found to be the major bioactive compounds in the *T. chebula* fruit samples.

Gallic acid, protocatechuic acid and quinic acid were in the ranges of 9,720–32,200, 505–2,370 and 13,600–22,200 µg/g, respectively. Ellagic acid was found in the range of 30,100–107,000 µg/g and detected maximum in TC7 and minimum in TC5. Catechin and epicatechin were found in the range of 49–3,070 and 245.5–5,520 µg/g, respectively. Catechin was not detected in TC2 and TC8.

Rutin was in the range of 59–1,180 µg/g, but not detected in TC2 and TC6. Eriodictyol, kaempferol and apigenin were found in the range of 24.15–563, 17.65–822 and 6.55–695 µg/g, respectively. Betulinic acid was detected only in TC1, TC5 and TC8. Arjunolic acid, arjunetin and arjungenin were found in the range of 1,490–4,470, 476–9,820 and 7,150–36,400 µg/g, respectively.

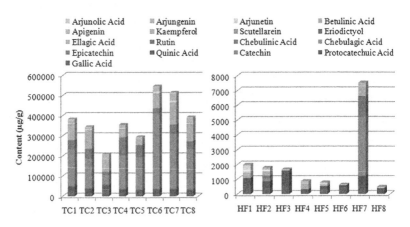

FIGURE 4.1   Graphical representation of major bioactive compounds detected in *Terminalia chebula* collected from different locations (coded as TC1–8) and its marketed polyherbal formulations (coded as HF1–8).

Significant variations were observed in the contents of the 17 analytes in the polyherbal formulations. The total content of the 17 bioactive compounds was found to be the highest in HF7 (7,526.45 µg/g) and lowest in HF8 (482.87 µg/g). Gallic acid, protocatechuic acid, catechin, quinic acid and epicatechin were found to be the most abundant compounds in HF1 (621 µg/g), HF3 (57.9 µg/g), HF8 (4.86 µg/g), HF3 (1,000 µg/g) and HF7 (2.06 µg/g), respectively. Chebulagic acid, chebulinic acid, betulinic acid, arjunetin, arjungenin and arjunolic acid were detected maximum in HF7 (3,990 µg/g), HF7 (1,420 µg/g), HF2 (70.5 µg/g), HF1 (484 µg/g), HF7 (175 µg/g) and HF4 (21.7 µg/g), respectively. Rutin, ellagic acid, eriodictyol, scutellarein, kaempferol and apigenin were detected maximum in HF7 (0.16 µg/g), HF7 (653 µg/g), HF8 (0.17 µg/g), HF5 (0.40 µg/g), HF4 (32.4 µg/g) and HF5 (1.99 µg/g), respectively. The data show wide variation in the contents in the herbal formulations showing a lack of standardization.

4.5 CONCLUSION

A rapid and sensitive UPLC-ESI-MS/MS method under MRM mode was successfully developed and validated for the simultaneous quantification of 17 bioactive phenolics and terpenoids in *T. chebula* fruits collected from different locations and its polyherbal formulations. A wide variation of the contents of the 17 bioactive compounds in *T. chebula* fruits from different locations indicated the effect of geographic location on the contents.

Results indicated that the total content of all the 17 compounds ranged from 54.6% w/w to 20.90% w/w. The samples from Lucknow and Jabalpur, on an average, showed almost similar total contents, whereas the samples from Kolkata showed 50% more content. Chebulagic acid showed the highest content in all fruit samples of *T. chebula* except for TC3. The polyherbal formulations investigated also showed wide variations and significant differences in the contents pointing out inadequate standardization. Comprehensive estimation of selected bioactive compounds could make a contribution to the quality control of *T. chebula* species and its herbal formulations.

References

Abdulkadir, A.R. "*In vitro* antioxidant activity of ethanolic extract from *Terminalia catappa* (L.) leaves and fruits: Effect of fruit ripening." *International Journal of Science and Research* 4, no. 8 (2013): 1244–1248.

Acharyya, S. and Prasenjit Bhuniya, A.S. "Evaluation of antimicrobial and anthelmintic activity of roots of *Terminalia paniculata.*" *The Pharma Innovation Journal* 8, no. 6 (2019): 1065–1068.

Adefegha, S.A., Oboh, G., Oyeleye, S.I. and Ejakpovi, I. "Erectogenic, antihypertensive, antidiabetic, anti-oxidative properties and phenolic compositions of almond fruit (*Terminalia catappa* L.) parts (Hull and Drupe) – *in vitro.*" *Journal of Food Biochemistry* 41, no. 2 (2017): e12309. doi:10.1111/jfbc.12309.

Agrawal, S., Kulkarni, G.T. and Sharma, V.N. "A comparative study on the antioxidant activity of methanolic extracts of *Terminalia paniculata* and Madhuca longifolia." *Free Radicals and Antioxidants* 1, no. 4 (2011): 62–68.

Ahirwar, B., Singhai, A.K. and Dixit, V.K. "Effect of *Terminalia chebula* fruits on lipid profiles of rats." *Journal of Natural Remedies* 3, no. 1 (2003): 31–35.

Alam, M.B., Zahan, R., Hasan, M., Khan, M.M., Rahman, M.S., Chowdhury, N.S. and Haque, M.E. "Thank you, a good research antioxidant, antimicrobial and toxicity studies of the different fractions of fruits of *Terminalia belerica* Roxb." *Global Journal of Pharmacology* 5, no. 1 (2011): 07–17.

Alladi, S., Prakash, S.D. and Nalini, M. "Antihyperglycemic activity of the leaves of *Terminalia tomentosa* against normal and alloxan induced diabetes rats." *Research Journal of Pharmacy and Technology* 5 (2012): 1573–1576.

Amalraj, A. and Gopi, S. "Medicinal properties of *Terminalia arjuna* (Roxb.) Wight & Arn.: A review." *Journal of Traditional and Complementary Medicine* 7 (2017): 65–78.

Anam, K., Widharna, R.M. and Kusrini, D. "α-Glucosidase inhibitor activity of Terminalia species." *International Journal of Pharmacology* 5, no. 4 (2009): 277–280.

Anand, A.V., Divya, N. and Kotti, P.P. "An updated review of *Terminalia catappa.*" *Pharmacognosy Reviews* 9 (2015): 93–98.

Anjaneyulu, A.S.R., Reddy, A.R., Mallavarapu, G.R. and Chandrasekhara, R.S. "3-Acetyl maslinic acid from the root bark of Terminalia alata." *Phytochemistry* 25 (1986): 2670–2671.

Arabind, K., Manivannan, E. and R. Chandrasekar, R. "Ethnopharmacological review of *Terminalia chebula.*" *Bioequiv. Bioequivalence & Bioavailability International Journal* 3, no. 1 (2019): 000135.

Arumugam, A. and Gopinath, K. "*In vitro* callus development of different explants used for different medium of *Terminalia arjuna.*" *Asian Journal of Biotechnology* 3, no. 6 (2011): 564–572.

Ashwini, R., Gajalakshmi, S., Mythili, S. and Sathiavelu, A. "*Terminalia chebula* – A pharmacological review." *Journal of Pharmacy Research* 4, no. 9 (2011): 2884–2887.

Asolkar, L.V., Kakkar, K.K. and Chakre, O.J. *Glossary of Indian Medicinal Plants with Active Principles*. New Delhi: Publications and Information's Directorate, CSIR, 1992: 230–231.

Avula, B., Wang, Y.H., Isaac, G., Yuk, J., Wrona, M., Yu, K. and Khan, I.A. "Metabolic Profiling of hoodia, chamomile, Terminalia Species and evaluation of commercial preparations using ultrahigh-performance liquid chromatography quadrupole-time-of-flight mass spectrometry." *Planta Medica* 83, no. 16 (2017): 1297–1308.

Avula, B., Wang, Y.H., Wang, M., Shen, Y.H. and Khan, I.A. Simultaneous determination and characterization of tannins and triterpene saponins from the fruits of various species of Terminalia and Phyllantus emblica using a UHPLC-UV-MS method: Application to triphala." *Planta Medica* 29, no. 2 (2013): 181–188.

Bag, A., Bhattacharyya, S.K. and Chattopadhyay, R.R. "The development of *Terminalia chebula* Retz. (Combretaceae) in clinical research." *Asian Pacific Journal of Tropical Biomedicine* 3, no. 3 (2013): 244–252.

Bagalkote, S., Hugar, S. and Javalgikar, A. "Evaluation of cardioprotective activity of *Terminalia catappa* leaves against doxorubicin induced myocardial infarction in albino rats." *International Journal of Current Medical and Pharmaceutical Research* 4, no. 2 (2018): 3045–3047.

Bajpai, M., Pande, A., Tewari, S.K. and Prakash, D. "Phenolic contents and antioxidant activity of some food and medicinal plants." *International Journal of Food Sciences and Nutrition* 56 (2005): 287–291.

Baliga, M.S., Meera, S., Mathai, B., Rai, M.P., Pawar, V. and Palatty, P.L. "Scientific validation of the ethnomedicinal properties of the Ayurvedic drug Triphala: A review." *Chinese Journal of Integrative Medicine* 18, no. 12 (2012): 946–954.

Barry, K.M., Davies, N.W. and Mohammed, C.L. "Identification of hydrolysable tannins in the reaction zone of *Eucalyptus nitens* wood by high performance liquid chromatography-electrospray ionisation mass spectrometry." *Phytochemical Analysis* 12 (2001): 120–127.

Beigi, M., Haghani, E., Alizadeh, A. and Samani, Z.N. "The pharmacological properties of several species of Terminalia in the world." *International Journal of Pharmaceutical Sciences and Research* 9, no. 10 (2018): 4079–4088.

Belapurkar, P., Goyal, P. and Tiwari-Barua, P. "Immunomodulatory Eeffects of Triphala and its individual constituents: A review." *Indian Journal of Pharmaceutical Sciences* 76, no. 6 (2014): 467–475.

Bharani, A., Ganguly, A. and Bhargava, K.D. "Salutory effect of *Terminalia arjuna* in patients with severe refractory heart failure." *International Journal of Cardiology* 49 (1995): 191–199.

Bodke, Y.D., Sindhe, M.A., Gupta, R.K. and Manjunatha, H. "Antioxidant and anthelmintic activity of *Terminalia arjuna* roxb. Stem bark extracts." *Asian Journal of Pharmaceutical and Clinical Research* 6, no. 4 (2013): 33–37.

Bystrom, L.M., Lewis, B.A., Brown, D.L., Rodriguez, E. and Obendorf, R.L. "Characterization of phenolics by LC-UV/vis, LC-MS/MS and sugars by GC in *Melicoccus bijugatus* Jacq. 'Montgomery' fruits." *Food Chemistry* 111 (2008): 1017–1024.

Callemien, D. and Collin, S. "Use of RP-HPLC-ESI(–)-MS/MS to differentiate various proanthocyanidin isomers in lager beer extracts." *Journal of the American Society of Brewing Chemists* 66 (2008): 109–115.

Chandra, S., Menpara, D. and Desai, D. "Antimicrobial activity of *Terminalia bellerica* leaf and stem collected from two different sites." *American Journal of Phytomedicine and Clinical Therapeutics* 1, no. 9 (2013): 721–733.

Chang, Z., Zhang, Q., Liang, W., Zhou, K., Jian, P., She, G. and Zhang, L. "A comprehensive review of the structure Elucidation of Tannins from Terminalia Linn." *Evidence-based Complementary and Alternative Medicine* 2019 (2019): 8623909, 26 pages. https://doi.org/10.1155/2019/8623909.

Charoenchai, L., Pathompak, P., Madaka, F., Settharaksa, S. and Saingam, W. "HPLC-MS profiles and quantitative analysis of triphala formulation." *Bulletin of Health Science and Technology* 14, no. 1 (2016): 57–67.

Chatha, S.A.S., Hussain, A.I., Asad, R., Majeed, M. and Aslam, N. "Bioactive components and antioxidant properties of *Terminalia arjuna* L. extracts." *Journal of Food Processing and Technology* 5 (2014): 298. doi:10.4172/2157-7110.1000298

Chattopadhyay, R.R. and Bhattacharyya, S. "*Terminalia chebula*: An update." *Pharmacognosy Reviews* 1, no. 1 (2007): 151–156.

Chaudhari, G.M. and Mahajan, R.T. "Comparative antioxidant activity of twenty traditional Indian medicinal plants and its correlation with total flavonoid and phenolic content." *International Journal of Pharmaceutical Sciences Review and Research* 30, no. 1 (2015): 105–111.

Chaudhari, M. and Mengi, S. "Evaluation of phytoconstituents of *Terminalia arjuna* for wound healing activity in rats." *Phytotherapy Research* 20 (2006): 799–805.

Chen, P.S., Li, J.H., Liu, T.Y. and Lin, T.C. "Folk medicine *Terminalia catappa* and its major tannin component, punicalagin, are effective against bleomycon-induced genotoxicity in Chinese hamster ovary cells." *Cancer Letters* 152, no. 2 (2000): 115–122.

Chen, Y., Zhou, G., Ma, B., Tong, J., and Wang, Y. "Active constituent in the ethyl acetate extract fraction of *Terminalia bellirica* fruit exhibits antioxidation, antifibrosis, and proapoptosis capabilities in vitro". *Oxidative Medicine and Cellular Longevity* 2019: 5176090, 15 pages. https://doi.org/10.1155/2019/5176090.

Cheng, H.Y., Lin, C.C. and Lin, T.C. "Antiherpes simplex virus type 2 activity of casuarinin from the bark of Terminalia arjuna Linn." *Antiviral Research* 55, no. 3 (2002): 447–455.

Cheng, H.Y., Lin, T.C., Yu, K.H., Yang, C.M. and Lin, C.C. "Antioxidant and free radical scavenging activities of *Terminalia chebula*." *Biological and Pharmaceutical Bulletin* 26, no. 9 (2003): 1331–1335.

Christenhusz, M.J. and Byng, J.W. "The number of known plants species in the world and its annual increase." *Phytotaxa* 261, no. 3 (2016): 201–217.

Cock, I.E. "The medicinal properties and phytochemistry of plants of the genus Terminalia (Combretaceae)." *Inflammopharmacology* 23 (2015): 203–229.

Corner, E.J.H. *Way Side Trees of Malaya* (Vol. 1, 4th edition). Kuala Lumpur: The Malayan Nature Society, 1997: 217.

Deb, A., Barua, S. and Das, B. "Pharmacological activities of Baheda (*Terminalia bellerica*): A review." *Journal of Pharmacognosy and Phytochemistry* 5, no. 1 (2016): 194–197.

Debnath, J., Prakash, T., Karki, R., Kotresha, D. and Sharma, P. "An experimental evaluation of anti-stress effects of *Terminalia chebula*." *Journal of Physiological and Biomedical Sciences* 24, no. 2 (2011): 13–19.

Devi, R.S., Kist, M., Vani, G. and Devi, C.S.S. "Effect of methanolic extract of *Terminalia arjuna* against Helicobacter pylori 26695 lipopolysaccharide-induced gastric ulcer in rats." *Journal of Pharmacy and Pharmacology* 60, no. 4 (2008): 505–514.

Dhanani, T., Shah, S. and Kumar, S. "A validated high-performance liquid chromatography method for determination of tannin-related marker constituents gallic acid, corilagin, chebulagic acid, ellagic acid and chebulinic Acid in four Terminalia species from India." *Journal of Chromatographic Science* 53, no. 4 (2015): 625–632.

Downeya, M.O. and Rochfort, S. "Simultaneous separation by reversed-phase high-performance liquid chromatography and mass spectral identification of anthocyanins and flavonols in Shiraz grape skin." *Journal of Chromatography A* 1201 (2008): 43–47.

Durge, A., Jadaun, P., Wadhwani, A., Chinchansure, A.A., Said, M., Thulasiram, H.V., Joshi, S.P. and Kulkarni, S.S. "Acetone and methanol fruit extracts of Terminalia paniculata inhibit HIV-1 infection in vitro." *Natural Product Research* 31, no. 12 (2017): 1468–1471.

Dwevedi, A., Dwivedi, R. and Sharma, Y.K. "Exploration of phytochemicals found in *Terminalia* sp. and their antiretroviral activities." *Pharmacognosy Review* 10, no. 20 (2016): 73–83.

Dwivedi, S. 2007. "*Terminalia arjuna* Wright & Arn. – A useful drug for cardiovascular disorders." *Journal of Ethnopharmacology* 114, no. 2 (2007): 114–129.

Dwivedi, S. and Chopra, D. "Revisiting *Terminalia arjuna* – An ancient cardiovascular drug." *Journal of Traditional and Complementary Medicine* 4, no. 4 (2014): 224–231.

Dwivedi, S. and Udupa, N. "*Terminalia arjuna*: Pharmacognosy, phytochemistry, pharmacology and clinical use. A review." *Fitoterpia* 60 (1989): 413–420.

Eesha, B.R., Mohanbabu, A.V., Meena, K.K., Vijay, M., Lalit, M. and Rajput, R. "Hepatoprotective activity of Terminalia paniculata against paracetamol induced hepatocellular damage in Wistar albino rats." *Asian Pacific Journal of Tropical Medicine* 4, no. 6 (2011): 466–469.

Ekambaram, S.P., Perumal, S.S. and Balakrishnan, A. "Tannin rich fraction from *Terminalia chebula* fruits as anti-inflammatory agent." *Journal of Herbs, Spices & Medicinal Plants* 24, no. 1 (2018): 74–86.

Eldeen, I.M., Elgorashi, E.E., Mulholland, D.A. and van Staden, J. "Anolignan B: A bioactive compound from the roots of *Terminalia sericea*." *Journal of Ethnopharmacology* 103 (2006): 135–138.

Elizabeth, K.M. "Antimicrobial activity of *Terminalia bellerica*." *Indian Journal of Clinical Biochemistry* 20, no. 2 (2005): 150–153.

Etienne, D.T., Christelle, K.K.A., Yves, N., Constant, K., Adama, C., Daouda, S., Ysidor, K.N.G. and Marius, B.G.H. "Antioxidants contents of *Terminalia catappa* (Combretaceae) almonds grown in Côte d'Ivoire." *Archives of Current Research International* 10 (2017): 1–12.

Fabre, N., Rustan, I., de Hoffmann, E. and Quetin-Leclercq, J. "Determination of flavone, flavonol, and flavanone aglycones by negative ion liquid chromatography electrospray ion trap mass spectrometry." *Journal of the American Society for Mass Spectrometry* 12, no. 6 (2001): 707–715.

Fahmy, N.M., Al-Sayed, E. and Singab, A.N. "Genus *Terminalia*: A phytochemical and biological review." *Medicinal and Aromatic Plants* 4 (2015): 1–22. doi:10.4172/2167-0412.1000218

Fan, Y.M., Xu, L.Z., Gao, J., Wang, Y., Tang, X.H., Zhao, X.N. and Zhang, Z.X. "Phytochemical and anti-inflammatory studies on *Terminalia catappa*." *Fitoterapia* 75 (2004): 253–260. doi:10.1016/j.fitote.2003.11.007

Fang, N., Yu, S. and Prior, R.L. "LC/MS/MS characterization of phenolic constituents in dried plums." *Journal of Agricultural and Food Chemistry* 50, no. 12 (2002): 3579–3585.

Farwick, M., Köhler, T., Schild, J., Mentel, M., Maczkiewitz, U., Pagani, V., Bonfigli, A., Rigano, L., Bureik, D. and Gauglitz, G.G. "Pentacyclic triterpenes from Terminalia arjuna show multiple benefits on aged and dry skin." *Skin Pharmacology and Physiology* 27 (2013): 71–81.

Gahlaut, A., Sharma, A., Shirolkar, A. and Dabur, R. "Non-targeted identification of compounds from regenerated bark of *Terminalia tomentosa* by HPLC- (+) ESI-QTOFMS." *Journal of Pharmacy Research* 6, no. 4 (2013): 415–418.

Gaikwad, D. and Jadhav, N. "A review on biogenic properties of stem bark of *Terminalia arjuna*: An update." *Asian Journal of Pharmaceutical and Clinical Research* 11, no. 8 (2018): 35–39.

Gandhi, N.M. and Nair, C.K.K. "Radiation protection by *Terminalia chebula* some mechanistic aspects." *Molecular and Cellular Biochemistry* 277, no. 1 (2005): 43–48.

Gangopadhyay, M. and Chakrabarty, T. "The family Combretaceae of Indian subcontinent." *Journal of Economic & Taxonomic Botany* 21, no (2) (1997): 281–365.

Ganjayi, M.S., Meriga, B., Hari, B., Oruganti, L., Dasari, S. and Mopuri, R. "Polyphenolic rich fraction of *Terminalia paniculata* attenuates obesity through inhibition of pancreatic amylase, lipase and 3T3-L1 adipocyte differentiation." *Journal of Nutrition & Intermediary Metabolism* 10 (2017): 19–25.

Ghosh, J., Das, J., Manna, P. and Sil, P.C. "Protective effect of the fruits of *Terminalia arjuna* against cadmium-induced oxidant stress and hepatic cell injury via MAPK activation and mitochondria dependent pathway." *Food Chemistry* 123 (2010): 1062–1075.

Hafiz, F.B., Towfique, N.M., Sen, M.K., Sima, S.N., Azhar, B.S. and Rahman, M.M. "A comprehensive ethno-pharmacological and phytochemical update review on medicinal plant of *Terminalia arjuna* Roxb. of Bangladesh." *Scholars Academic Journal of Pharmacy* 3, no. 1 (2014): 19–25.

Hager, J.W. and Le Blanc, J.Y. "High-performance liquid chromatography-tandem mass spectrometry with a new quadrupole/linear ion trap instrument." *Journal of Chromatography A* 1020, no. 1 (2003): 3–9.

Halder, S., Bharal, N., Mediratta, P.K., Kaur, I. and Sharma, K.K. "Antiinflammatory, immunomodulatory and antinociceptive activity of Terminalia arjuna Roxb bark powder in mice and rats." *Indian Journal of Experimental Biology* 47 (2009): 577–583.

Hedina, A., Kotti, P., Kausar, J. and Anand, V. "Phytopharmacological overview of *Terminalia chebula* Retz." *Pharmacognosy Journal* 8, no. 4 (2016): 307–309.

Hemalatha, T., Pulavendran, S., Balachandran, C., Manohar, B.M. and Puvana-krishnan, R. "Arjunolic acid: A novel phytomedicine with multifunctional thera-peutic applications." *Indian Journal of Experimental Biology* 48, no. 3 (2010): 238–247.

Hordyjewska, A., Ostapiuk, A., Horecka, A. and Kurzepa, J. "Betulin and betu-linic acid: Triterpenoids derivatives with a powerful biological potential." *Phytochemistry Reviews* 18 (2019): 929–951.

ICH Guideline. Q2 (R1), Validation of analytical procedures: Text and methodology. *International Conference on Harmonization*, Geneva, 2005, p. 1.

Israni, D.A., Patel, K.V. and Gandhi, T.R. "Anti-hyperlipidemic activity of aqueous extract of Terminalia chebula and Gaumutra in high cholesterol diet fed rats." *Pharma Science Monitor-An International Journal Pharmaceutical Sciences* 1, no. 1 (2010): 48–59.

Jadon, A., Bhadauria, M. and Shukla, S. "Protective effect of *Terminalia belerica* Roxb. and gallic acid against carbon tetrachloride induced damage in albino rats." *Journal of Ethnopharmacology* 109 (2007): 214–218.

Jain, S., Yadav, P.P., Gill, V., Vasudeva, N. and Singla, N. "*Terminalia arjuna* a sacred medicinal plant: Phytochemical and pharmacological profile." *Phytochemistry Review* 8 (2009): 491–502.

Jaiswal, R., Jayasinghe, L. and Kuhnert, N. "Identification and characterization of pro-anthocyanidins of 16 members of the Rhododendrongenus (Ericaceae) by tan-dem LC–MS." *Journal of Mass Spectrometry* 47 (2012): 502–515.

Jensi, V.D. and Gopu, P.A. "Evaluation of hypolipidemic activity of various phytocon-stituents from *Terminalia arjuna* (Roxb. ex DC.) in rat fed with high fat diet." *International Journal of Pharmaceutical Sciences Review and Research* 50, no. 2 (2018): 41–48.

Jitta, S.R., Daram, P., Gourishetti, K., Misra, C.S., Polu, P.R., Shah, A., Shreedhara, C.S., Nampoothiri, M. and Lobo, R. "*Terminalia tomentosa* bark amelio-rates inflammation and arthritis in carrageenan induced inflammatory model and Freund's Adjuvant-induced arthritis model in rats." *Journal of Toxicology* (2019): 7898914, 11 pages. https://doi.org/10.1155/2019/7898914.

Jokar, A., Masoomi, F., Sadeghpour, O., Nassiri-Toosi, M. and Hamedi, S. "Potential therapeutic applications for *Terminalia chebula* in Iranian traditional medicine." *Journal of Traditional Chinese Medicine* 36, no. 2 (2016): 250–254.

Joshi, A.B., Aswathi, M. and Bhobe, M. "*Terminalia tomentosa* Roxb (ex DC) Wight & Arn: Phytochemical Investigation." *American Journal of Advanced Drug Delivery* 1, no. 3 (2013): 224–231.

Joshi, V.K., Joshi, A. and Dhiman, K.S. "The Ayurvedic Pharmacopoeia of India, development and perspectives." *Journal of Ethnopharmacology* 197 (2017): 32–38.

Juang, L.J., Sheu, S.J. and Lin, T.C. "Determination of hydrolyzable tannins in the fruit of *Terminalia chebula* Retz. by high-performance liquid chromatography and capillary electrophoresis." *Separation Science* 27, no. 9 (2004): 718–724.

Kalola, J. and Rajani, M. "Extraction and TLC desitometric determination of triterpe-noid acids (arjungenin, arjunolic acid) from Terminalia arjuna stembark without interference of tannins." *Chromatographia* 63 (2006): 475–481.

Kamboj, V.P. "Herbal medicine." *Current Science* 78 (2000): 35–39.

Kaur, S. and Jaggi, R.K. "Antinociceptive activity of chronic administration of different extracts of *Terminalia bellerica* Roxb. and *Terminalia chebula* Retz. Fruits" *Indian Journal of Experimental Biology* 48 (2010): 925–930.

Khaliq, F. and Fahim, M. "Role of *Terminalia Arjuna* in improving cardiovascular functions: A review." *Indian Journal of Physiology and Pharmacology* 62, no. 1 (2018): 8–19.

Khan, A.U. and Gilani, A.H. "Pharmacodynamic evaluation of *Terminalia belerica* for its anti hypertensive effect." *Journal of Food and Drug Analysis* 16 (2008): 6–14.

Khan, M.S.A., Hasan, M.W., Shereen, M., Sultana, T., Dastagir, I.M., Ali, A.J., Qureshi, S., Ghori, S.S. and Hussain, S.A. "Anti-nociceptive effect of *Terminalia coriacea* (Roxb.) Wight and Arn. leaf methanolic extract." *PharmacologyOnLine* 7 (2011): 1176–1189.

Khan, M.S.A., Khatoon, N., Al-Sanea, M.M., Mahmoud, M.G. and Rahman, H.U. "Methanolic extract of Leathery Murdah, *Terminalia coriacea* (Roxb.) Wight and Arn. leaves exhibits anti-inflammatory activity in acute andc models." *Medical Principles and Practice* 27, no. 3 (2018): 267–271.

Khan, A.A., Kumar, V., Singh, B.K. and Singh, R. "Evaluation of wound healing property of *Terminalia catappa* on excision wound models in Wistar rats." *Drug Research (Stuttg)* 64, no. 5 (2014): 225–228.

Khan, M.S.A., Mat Jais, A.M., Zakaria, Z.A., Mohtarrudin, N., Ranjbar, M. and Khan, M. "Wound healing potential of Leathery Murdah, *Terminalia coriacea* (Roxb.) Wight and Arn." *Phytopharmacology* 3 (2012): 158–168.

Khan, M.S.A., Nazan, S. and Jais, A.M. "Flavonoids and anti-oxidant activity mediated gastroprotective action of Leathery Murdah, *Terminalia coriacea* (Roxb.) Wight and Arn. leaf methanolic extract in rats." *Arquivos de Gastroenterologia* 54, no. 3 (2017): 183–191.

Khare, C.P. *Encyclopedia of Indian Medicinal Plants.* Berlin: Springer-Verlag, 2004: 451–453.

Khare, C.P. *The Indian Medicinal Plants – An Illustrated Dictionary.* New York: Springer-Verlag, 2007: 655–656.

Khatoon, S., Singh, N., Srivastava, N., Rawat, A. and Mehrotra, S. "Chemical evaluation of seven Terminalia species and quantification of important polyphenols by TLC." *Journal of Planar Chromatography – Modern TLC* 21, no. 3 (2008): 167–171.

Kim, M.S., Lee, D.Y., Sung, S.H. and Jeon, W.K. "Anti-cholinesterase activities of hydrolysable tannins and polyhydroxy triterpenoid derivatives from *Terminalia chebula* Retz. fruit." *Records of Natural Products* 12, no. 3 (2018): 284–289.

Kinoshita, S., Inoue, Y., Nakama, S., Ichiba, T. and Aniya, Y. "Antioxidant and hepatoprotective actions of medicinal herb, *Terminalia catappa* L. from Okinawa Island and its tannin corilagin." *Phytomedicine* 14, no. 11 (2007): 755–762.

Kirtikar, K.R and Basu, B.D. *Indian Medicinal Plants.* Allahabad: L M Basu Publication, 1989.

Kirtikar, K.R. and Basu, B.D. *Indian Medicinal Plants* (Vols. I–IV). Allahabad: Lalit Mohan Basu, 1935.

Kirtikar, K.R. and Basu, B.D. *Indian Medicinal Plants.* Delhi: Periodical Expert's Book Agency, 1991: 1016, 1028.

Kong, W., Wen, J., Yang, Y., Qiu, F., Sheng, P. and Yang, M. "Simultaneous targeted analysis of five active compounds in licorice by ultra-fast liquid chromatography coupled to hybrid linear-ion trap tandem mass spectrometry." *Analyst* 139, no. 8 (2014): 1883–1894.

Krishnaveni, M. "In vitro antioxidant activity of *Terminalia catappa* nuts." *Asian Journal of Pharmaceutical and Clinical Research* 7, no. 4 (2014): 33–35.

Kumar, K.J. "Effect of geographical variation on contents of tannic acid, gallic acid, chebulinic acid and ethyl gallate in *Terminalia chebula*." *Natural Products an Indian Journal* 2, no. 3–4 (2006): 100–104.

Kumar, N. "Phytopharmacological overview on *Terminalia arjuna* Wight and Arn." *World Journal of Pharmaceutical Sciences* 2, no. 11 (2014): 1557–1566.

Kumar, S., Dobos, G.J. and Rampp, T. "The significance of ayurvedic medicinal plants." *Journal of Evidence-Based Complementary & Alternative Medicine* 22, no. 3 (2017): 494–501.

Kumar, N. and Khurana, S.M. "Phytochemistry and medicinal potential of the *Terminalia bellirica* Roxb (Bahera)." *Indian Journal of Natural Products and Resources* 9, no. 2 (2018): 97–107.

Kumar, A., Lakshman, K., Jayaveera, K.N., Tripathi, S.M. and Satish, K.V. "Estimation of gallic acid, rutin and quercetin in *Terminalia chebula* by HPTLC." *Jordan Journal of Pharmaceutical Sciences* 3, no. 1 (2010): 63–67.

Kumari, S., Mythili Krishna, J., Joshi, A.B., Gurav, S., Bhandarkar, A.V., Agarwal, A., Deepak, M. and Gururaj, G.M. "A pharmacognostic, phytochemical and pharmacological review of *Terminalia bellerica*." *Journal of Pharmacognosy and Phytochemistry* 6, no. 5 (2017): 368–376.

Latha, R.C.R. and Daisy, P. "Insulin-secretagogue, antihyperlipidemic and other protective effects of gallic acid isolated from *Terminalia bellerica* Roxb. in streptozotocin-induced diabetic rats." *Chemico Biological Interactions* 189, no. 1–2 (2011): 112–118.

Lee, D., Boo, K.H., Woo, J.K., Duan, F., Lee, K.H., Kwon, T.K., Lee, H.Y., Riu, K.Z. and Lee, D.S. "Anti-bacterial and anti-viral activities of extracts from *Terminalia chebula* barks." *Journal of the Korean Society for Applied Biological Chemistry* 52, no. 2 (2011): 295–298.

Lee, H.S., Jung, S.H., Yun, B.S. and Lee, K.W. "Isolation of chebulic acid from *Terminalia chebula* Retz. and its antioxidant effect in isolated rat hepatocytes." *Archives of Toxicology* 81, no. 3 (2007): 211–218.

Li, H.J. and Deinzer, M.L. "Tandem mass spectrometry for sequencing proanthocyanidins." *Analytical Chemistry* 79, no. 4 (2007): 1739–1748.

Li, K., Diao, Y., Zhang, H., Wang, S., Zhang, Z., Yu, B., Huang, S. and Yang, H. "Tannin extracts from immature fruits of Terminalia chebula Fructus Retz. promote cutaneous wound healing in rats." *BMC Complementary and Alternative Medicine* 11, no. 1 (2011): 86.

Li, S., Ye, T., Liang, L., Liang, W., Jian, P., Zhou, K. and Zhang, L. "Anti-cancer activity of an ethyl-acetate extract of the fruits of Terminalia bellerica (Gaertn.) Roxb. through an apoptotic signaling pathway in vitro." *Journal of Traditional Chinese Medical Sciences* 5, no. 4 (2018): 370–379.

Li, D.Q., Zhao, J., Xie, J. and Li, S.P. "A novel sample preparation and on-line HPLC-DAD-MS/MS-BCD analysis for rapid screening and characterization of specific enzyme inhibitors in herbal extracts: Case study of α-glucosidase." *Journal of Pharmaceutical and Biomedical Analysis* 88 (2014): 130–135.

Lin, T.C. "Study on the tannins and related compounds in the fruit of *Terminalia catappa* L." *Journal of Chinese Medical and Pharmaceutical Research* 14 (1992): 165–174.

Lin, C.C., Hsu, Y.F. and Lin, T.C. "Effects of punicalagin and punicalin on carrageenan-induced inflammation in rats." *The American Journal of Chinese Medicine* 27, no. 3–4 (1999): 371–376.

Lin, Y.L., Kuo, Y.H., Shiao, M.S., Chen, C.C. and Ou, J.C. "Flavonoid glycosides from *Terminalia catappa* L." *Jornal of the Chinese Chemical Society* 47, no. 1 (2000): 253–256.

Lin, Y., Wu, B., Li, Z., Hong, T., Chen, M., Tan, Y., Jiang, J. and Huang, C. "Metabolite identification of myricetin in rats using HPLC coupled with ESI-MS." *Chromatographia* 75 (2012): 655–660.

Mahajan, A. and Pai, N. "Simultaneous isolation and identification of phytoconstituents from *Terminalia chebula* by preparative chromatography." *Journal of Chemical and Pharmaceutical Research* 2, no. 5 (2010): 97–103.

Mahato, S.B., Nandy, A.K. and Kundu, A.P. "Pentacyclic triterpenoid sapogenols and their glycosides from *Terminalia bellerica*." *Tetrahedron* 48, no. 12 (1992): 2483–2494.

Mahesh, R., Bhuvana, S. and Hazeena Begum, V.M. "Effect of *Terminalia chebula* aqueous extract on oxidative stress and anti-oxidant status in the liver and kidney of young and aged rats." *Cell Biochemistry and Function* 27, no. 6 (2009): 358–363.

Malekzadeh, F., Ehsanifar, H., Shahamat, M., Levin, M. and Colwell, R.R. "Antibacterial activity of black myrobalan (Terminalia chebula Retz) against Helicobacter pylori." *International Journal of Antimicrobial Agents* 18, no. 1 (2001): 85–88.

Mallavarapu, G.R., Rao, S.B., Muralikrishna, E. and Rao, G.S. "Triterpenoids of the heartwood of *Terminalia alata* Heyne ex Roth." *Indian Journal of Chemistry – Section B: Organic and Medicinal Chemistry* 19, no. 8 (1980): 713–714.

Mallavarapu, G.R., Rao, S.B. and Syamasundar, K.V. "Chemical constituents of the bark of *Terminalia alata*." *Journal of Natural Products* 49, no. 3 (1986): 549–550.

Mallik, J., Al, F.A. and Kumar, B.R. "A Comprehensive review on pharmacological activity of *Terminalia catappa* (Combretaceae) – An update." *Asian Journal of Pharmaceutical Research and Development* 1, no. 2 (2013): 65–70.

Mandal, S., Patra, A., Samanta, A., Roy, S., Mandal, A., Mahapatra, T.D., Pradhan, S., Das, K. and Nandi, D.K. "Analysis of phytochemical profile of *Terminalia arjuna* bark extract with antioxidative and antimicrobial properties." *Asian Pacific Journal of Tropical Biomedicine* 3, no. 12 (2013): 960–966.

Mandloi, S., Mishra, R., Varma, R., Varughese, B. and Tripathi, J. "A study on phytochemical and antifungal activity of leaf extracts of *Terminalia cattapa*." *International Journal of Pharma and Bio Sciences* 4, no. 4 (2013b): 1385–1393.

Mandloi, S., Srinivasa, R., Mishra, R. and Varma, R. "Antifungal Activity of Alcoholic Leaf Extracts of *Terminalia catappa* and *Terminalia arjuna* on some pathogenic and allergenic fungi." *Advances in Life Science and Technology* 8 (2013a): 25–27.

Manipal, K., Lagisetty, R. and Chetty, K.M. Quantitation of flavonoids in barks of selected taxa of combretaceae. *Pharmacy & Pharmacology International Journal* 5, no. 1 (2017): 26–29.

Manna, P., Sinha, M., Pal, P. and Sil, P.C. "Arjunolic acid, a triterpenoid saponin, ameliorates arsenic-induced cyto-toxicity in hepatocytes." *Chemico-Biological Interactions* 170, no. 3 (2007b): 187–200.

Manna, P., Sinha, M. and Sil, P.C. "Phytomedicinal activity of *Terminalia arjuna* against carbon tetrachloride induced cardiac oxidative stress." *Pathophysiology* 14 (2007a): 71–78.

Mard, S.A., Veisi, A., Naseri, M.K.G. and Mikaili, P. "Spasmogenic activity of the seed of *Terminalia chebula* Retz in rat small intestine: *In vivo* and *in vitro* studies." *Malaysian Journal of Medical Sciences* 18, no. 3 (2011): 18–26.

MassBank Record: PT202000 https://massbank.eu/MassBank/RecordDisplay.jsp?id=PT202000&dsn=RIKEN_ReSpect

MassBank Record: BML00601 http://www.massbank.jp/jsp/Dispatcher.jsp?type=disp&id=BML00601&site=22

MassBank Record: BML00729 http://www.massbank.jp/jsp/Dispatcher.jsp?type=disp&id=BML00729&site=22

MassBank Record: BML01698. http://massbank.normandata.eu/MassBank/jsp/Dispatcher.jsp?type=disp&id=BML0169 8&site=25

MassBank Record: BML82316 http://www.massbank.jp/jsp/Dispatcher.jsp?type=disp&id=BML82316&site=22

MassBank Record: FIO00624 http://www.massbank.jp/jsp/Dispatcher.jsp?type=disp&id=FIO00624&site=25

MassBank Record: FIO00705 http://www.massbank.jp/jsp/Dispatcher.jsp?type=disp&id=FIO00705&site=25

MassBank Record: KOX00321, http://massbank.normandata.eu/MassBank/jsp/Dispatcher.jsp?type=disp&id=KOX0032 1&site=7

MassBank Record: PR100485, http://massbank.normandata.eu/MassBank/jsp/Dispatcher.jsp?type=disp&id=PR100485 &site=5.

MassBank Record: PR100984 http://www.massbank.jp/jsp/Dispatcher.jsp?type=disp&id=PR100984&site=1

McGaw, L.J., Rabe, T., Sparg, S.G., Jäger, A.K., Eloff, J.N. and Van Staden, J. "An investigation on the biological activity of Combretum spp." *Journal of Ethnopharmacology* 75, no. 1 (2001): 45–50.

Meena, A.K., Yadav, A., Singh, U., Singh, B., Sandeep, K. and Rao, M.M. "Evaluation of physicochemical parameters on the fruit of *Terminalia bellirica* Roxb." *International Journal of Pharmacy and Pharmaceutical Sciences* 2, no. S2 (2010): 96–98.

Mena, P., Calani, L., Dall'Asta, C., Galaverna, G., García-Viguera, C., Bruni, R., Crozier, A. and Del Rio, D. "Rapid and comprehensive evaluation of (poly)phenolic compounds in pomegranate (*Punica granatum* L.) juice by UHPLC-MS^n." *Molecules* 17, no. 12 (2012): 14821–14840.

Meriga, B., Naidu, P.B., Muniswamy, G., Kumar, G.H., Naik, R.R. and Pothani, S. "Ethanolic fraction of *Terminalia tomentosa* attenuates biochemical and physiological derangements in diet induced obese rat model by regulating key lipid metabolizing enzymes and adipokines." *Pharmacognosy Magazine* 13, no. 51 (2017): 385–392.

Mininel, F.J., Leonardo Junior, C.S., Espanha, L.G., Resende, F.A., Varanda, E.A., Leite, C.Q.F., Vilegas, W. and Dos Santos, L.C. "Characterization and quantification of compounds in the hydroalcoholic extract of the leaves from *Terminalia catappa* Linn. (Combretaceae) and their mutagenic activity." *Evidence-Based Complementary and Alternative Medicine* 2014 (2014): 676902, 11 pages.

Mishra, V., Agrawal, M., Onasanwo, S.A., Madhur, G., Rastogi, P., Pandey, H.P., Palit, G. and Narender, T. "Anti-secretory and cyto-protective effects of chebulinic acid isolated from the fruits of Terminalia chebula on gastric ulcers." *Phytomedicine* 20, no. 6 (2013): 506–511.

Modak, M., Dixit, P., Londhe, J., Ghaskadbi, S. and Devasagayam, T.P.A. "Indian herbs and herbal drugs used for the treatment of diabetes." *Journal of Clinical Biochemistry and Nutrition* 40, no. 3 (2007): 163–173.

Moghaddam, M.G., Ahmad, F.B.H. and Samzadeh-Kermani, A. "Biological activity of betulinic acid: A review." *Pharmacology and Pharmacy* 3, no. 2 (2012): 119–123.

Monnet, Y.T., Gbogouri, A., Koffi, P.K.B. and Kouamé, L.P. " Chemical characterization of seeds and seed oils from mature *Terminalia catappa* fruits harvested in Côte d'Ivoire." *International Journal of Biosciences* 10, no. 1 (2012): 110–124.

Mopuri, R., Ganjayi, M., Banavathy, K.S., Parim, B.N. and Meriga, B. "Evaluation of anti-obesity activities of ethanolic extract of *Terminalia paniculata* bark on high fat diet-induced obese rats." *BMC Complementary and Alternative Medicine* 15 (2015): 76. doi:10.13140/RG.2.1.4422.0249.

Mopuri, R. and Meriga, B. "In vitro anti oxidant activity and acute oral toxicity of *Terminalia paniculata* bark ethanolic extract on Sprague Dawley rats." *Asian Pacific Journal of Tropical Biomedicine* 4, no. 4 (2014): 294–298.

Mostafa, M.G., Rahman, M. and Karim, M.M. "Antimicrobial activity of *Terminalia chebula*." *International Journal of Medicinal and Aromatic Plants* 1, no. 2 (2011): 175–179.

Mukherjee, P.K. "Problems and prospects for good manufacturing practice for herbal drugs in Indian systems of medicine." *Drug Information Journal* 36, no. 3 (2002): 635–644.

Mukherjee, P.K., Rai, S., Bhattacharya, S., Wahile, A. and Saha, B.P. "Marker analysis of polyherbal formulation, Triphala – A well known Indian traditional medicine." *Indian Journal of Traditional Knowledge* 7, no. 3 (2008): 379–383.

Murali, Y.K., Chandra, R. and Murthy, P.S. "Antihyperglycemic effect of water extract of dry fruits of *Terminalia chebula* in experimental diabetes mellitus." *Indian Journal of Clinical Biochemistry* 19, no. 2 (2004): 202–204.

Nadkarni, A.K. "*Terminalia paniculata Roxb.*" In *Dr KM Nadkarni's Indian Materia Medica* (Vol. 1, 3rd edition). K. M. Nadkarni, Mumbai: Popular Prakashan, 1996: 931.

Nagappa, A.N., Thakurdesai, P.A., Rao, N.V. and Singh, J. "Antidiabetic activity of *Terminalia catappa* Linn fruits." *Journal of Ethnopharmacology* 88 (2003): 45–50.

Nair, R. and Chanda, S. "Antimicrobial activity of *Terminalia catappa*, Manilkarazapota and *Piper betel* leaf extract." *Indian Journal of Pharmaceutical Sciences* 70, no. 3 (2008): 390–393.

Nair, V., Singh, S. and Gupta, Y.K. "Anti-arthritic and disease modifying activity of *Terminalia chebula* Retz. in experimental models." *Journal Pharmacy and Pharmacology* 62, no. 12 (2010): 1801–1806.

Nampoothiri, S.V., Prathapan, A., Cherian, O.L., Raghu, K.G., Venugopalan, V.V. and Sundaresan, A. In vitro antioxidant and inhibitory potential of *Terminalia bellerica* and *Emblica officinalis* fruits against LDL oxidation and key enzymes linked to type 2 diabetes. *Food and Chemical Toxicology* 49, no. 1 (2011): 125–131.

Nandy, A.K., Podder, G., Sahu, N.P. and Mahato, S.B. "Triterpenoids and their gluco-sides from *Terminalia bellerica.*" *Phytochemistry* 28, no. 10 (1989): 2769–2772.

Narayan, C.L. and Rai, R.V. "Anti-HIV-1 Activity of Ellagic acid Isolated from *Terminalia paniculata.*" *Free Radicals and Antioxidants* 6, no. 1 (2016): 101–108.

Nguyen, Q.V., Eun, J.B., Wang, S.L., Nguyen, D.H., Tran, T.N. and Nguyen, A.D. "Anti-oxidant and Antidiabetic effect of some medicinal plants belong to Terminalia species collected in Dak Lak province, Vietnam." *Research on Chemical Intermediates* 42 (2016): 5859–5871.

Nunes, A.F., Sousa, P.V.L., Viana, V.S.L., Brito-Jr, E.C., Rabelo, R.S., Nunes-Filho, D.M., Nunes, P.H.M. and Martins, M.C. "Antiulcerogenic activity of ethanol extract of the bark from *Terminalia catappa* in gastric ulcer model induced by ethanol in Rattus norvegicus." *Pharmacologyonline* 1 (2012): 98–101.

Opara, F.N., Anuforo, H.U., Akujobi, C.O., Adjero, A., Mgbemena, I.C. and Okechuk Wu, R.I. "Preliminary phytochemical screening and antibacterial activities of leaf extracts of *Terminalia catappa.*" *Journal of Emerging Trends in Engineering and Applied Sciences* 3, no. 3 (2012): 424–428.

Paarakh, P.M. "*Terminalia arjuna* (Roxb.) Wt. and Arn.: A Review." *International Journal of Pharmacology* 6 (2010): 515–534.

Pan, S.Y., Litscher, G., Gao, S.H., Zhou, S.F., Yu, Z.L., Chen, H.Q., Zhang, S.F., Tang, M.K., Sun, J.N. and Ko, K.M. "Historical perspective of traditional indig-enous medical practices: The current renaissance and conservation of herbal resources." *Evidence-based Complementary and Alternative Medicine* (2014): Article ID 525340.

Parrotta, J. *Healing Plants of Peninsular India.* Oxford and New York: CABI Publishing, 2001.

Pasha, S.G., Khateeb, M.S., Pasha, S.A., Khan, M.S.A. and Shankaraiah, P. "Anti- epi-leptic activity of methanolic extract of *Terminalia coriacea* (Roxb.) Wight & Arn. in rats." *Journal of Advanced Pharmaceutical Sciences* 3, no. 2 (2013): 502–510.

Pasha, S.G., Shamshudin, S.K., Ahamed, S.Y., Imaduddin, M.D., Roshan, S. and Khan, M.S.A. "Evaluation of antiepileptic and antioxidant activity of metha-nolic extract of *Terminalia tomentosa* (ROXB) wight and ARN in rats." *World Journal of Pharmaceutical Research* 4, no. 2 (2014): 766–776.

Patel, J., Reddy, A.V., Kumar, G.S., Bajari, D.S.B. and Nagarjuna, V. "Hepato- pro-tective activity of methanolic extract of *Terminalia coriacea* leaves." *Research Journal of Pharmacy and Technology* 10, no. 5 (2017): 1313–1316.

Pettit, G.R., Hoard, M.S., Doubek, D.L., Schmidt, J.M., Pettit, R.K., Tackett, L.P. and Chapuis, J.C. "Antineoplastic agents 338. The cancer cell growth inhibitory. Constituents of *Terminalia arjuna* (Combretaceae)." *Journal of Ethnopharmacology* 53, no. 2 (1996): 57–63.

Pfundstein, B., El Desouky, S.K., Hull, W.E., Haubner, R., Erben, G. and Owen, R.W. "Polyphenolic compounds in the fruits of Egyptian medicinal plants (*Terminalia bellerica, Terminalia chebula and Terminalia horrida*): Characterization, quan-titation and determination of antioxidant capacities." *Phytochemistry* 71, no. 10 (2010): 1132–1148.

Prakash, J., Srivastava, S., Ray, R.S., Singh, N., Rajpali, R. and Singh, G.N. "Current status of herbal drug standards in the Indian Pharmacopoeia." *Phytotherapy Research* 31, no. 12 (2017): 1817–1823.

Promila, P. and Madan, V.K. Therapeutic and phytochemical profiling of *Terminalia chebula* Retz. (Harad): A review. *Journal of Medicinal Plants Studies* 6, no. 2 (2018): 25–31.

Raju, V.S. and Reddy, K.N. "Ethnomedicine for dysentery and diarrhea from Khammam district of Andhra Pradesh." *Indian Journal of Traditional Knowledge* 4, no. 4 (2005): 443–447.

Ram, A., Lauria, P., Gupta, R., Kumar, P. and Sharma, V.N. "Hypocholesterolaemic effects of *Terminalia arjuna* tree bark." *Journal of Ethnopharmacology* 55, no. 3 (1997): 165–169.

Ramachandran, S., Rajasekaran, A. and Adhirajan, N. "In Vivo and In Vitro Antidiabetic Activity of *Terminalia paniculata* Bark: An evaluation of possible phytoconstituents and mechanisms for blood glucose control in diabetes." *International Scholarly Research Notices Pharmacology* 2013 (2013): Article ID 484675, 10 pages.

Ramachandran, S., Rajasekaran, A. and Manisenthilkumar, K.T. "Investigation of hypoglycemic, hypolipidemic and antioxidant activities of aqueous extract of *Terminalia paniculata* bark in diabetic rats." *Asian Pacific Journal of Tropical Biomedicine* 2, no. 4 (2012): 262–268.

Rane, S., Hate, M., Hande, P. and Datar, A. "Development and validation of LC- MS/MS method for the simultaneous determination of arjunic acid, arjungenin and arjunetin in *Terminalia arjuna* (Roxb.) Wight & Arn." *International Journal of Science and Research* 5, no. 11 (2016): 1097–1001.

Rao, N.K. and Nammi, S. "Anti-diabetic and renoprotective effects of the chloroform extract of *Terminalia chebula* seeds in streptozotocin-induced diabetic rats." *BMC Complementary and Alternative Medicine* 6 (2006): 17.

Rathinamoorthy, R. and Thilagavathi, G. "*Terminalia chebula* – Review on pharmacological and biochemical studies." *International Journal of Pharm Tech Research* 6, no. 1 (2014): 97–116.

Reddy, D.B., Reddy, T.C.M., Jyotsna, G., Sharan, S., Priya, N., Lakshmipathi, V. and Reddanna, P. "Chebulagic acid, a COX-LOX dual inhibitor isolated from the fruits of *Terminalia chebula* Retz. induces apoptosis in COLO-205 cell line." *Journal of. Enthnopharmacology* 124, no. 3 (2009): 506–512.

Riaz, M., Khan, O., Sherkheli, M.A., Khan, M.Q. and Rashid, R. "Chemical constituents of *Terminalia chebula.*" *Natural Products Indian Journal* 13, no. 2 (2017): 112–127.

Row, L.R. and Raju, R.R. "Chemistry of terminalia species—XI: Isolation of 3,4,3′-O-trimethyl flavellagic acid from *Terminalia paniculata* roth." *Tetrahedron* 23, no. 2 (1967): 879–884.

Row, L.R. and Rao, G.S. "Chemistry of Terminalia species-VI: The constitution of tomentosic acid, a new triterpene carboxylic acid from *Terminalia tomentosa* wight et arn." *Tetrahedron* 18, no. 7 (1962a): 827–838.

Row, L.R. and Rao, G.S. "Chemistry of Terminalia species—III Chemical examination of *Terminalia paniculata* roth." *Tetrahedron* 18, no. 3 (1962b): 357–360.

Rumalla, C.S., Avula, B., Ali, Z., Wang, W., Smillie, T.J. and Khan, I.A. "Quantitative determination of saponins in *Terminalia chebula* and comparative study of Terminalia species by High Performance Thin Layer Chromatography." *Planta Medica* 76, no. 5 (2010): P32.

Saha, P.K., Patrab, P.H., Pradhan, R., Radharaman Dey, S.D. and Mandal, T.K.V. "Effect of *terminalia chebula* and *terminalia belerica* on wound healing in induced dermal wounds in rabbits." *Pharmacologyonline* 2 (2011): 235–241.

Saha, A., Pawar, V.M. and Jayaraman, S. "Characterisation of polyphenols in *Terminalia arjuna* bark extract." *Indian Journal of Pharmaceutical Sciences* 74, no. 4 (2012): 339–347.

Saleem, A., Husheem, M., Härkönen, P. and Pihlaja, K. "Inhibition of cancer cell growth by crude extract and the phenolics of *Terminalia chebula* Retz. fruit." *Journal of Ethnopharmacology* 81, no. 3 (2002): 327–336.

Sánchez-Rabaneda, F., Jáuregui, O., Casals, I., Andrés-Lacueva, C., Izquierdo-Pulido, M. and Lamuela-Raventós, R.M. "Liquid chromatographic/electrospray ionization tandem mass spectrometric study of the phenolic composition of cocoa (*Theobroma cacao*)." *Journal of Mass Spectrometry* 38, no. 1 (2003): 35–42.

Sandhu, J.S., Shah, B., Shenoy, S., Chauhan, S., Lavekar, G.S. and Padhi, M.M. "Effects of *Withania somnifera* (Ashwagandha) and *Terminalia arjuna* (Arjuna) on physical performance and cardiorespiratory endurance in healthy young adults." *International Journal of Ayurveda Research* 1, no. 3 (2010): 144–149.

Sangavi, R., Venkatalakshmi, P. and Brindha, P. "Anti-bacterial activity of *Terminalia catappa* L. bark against some bacterial pathogens." *World Journal of Pharmacy and Pharmaceutical Sciences* 4, no. 9 (2015): 987–992.

Satardekar, K.V. and Deodhar, M. "Anti-aging Ability of Terminalia species with special reference to hyaluronidase, elastase inhibition and collagen synthesis *in vitro*." *International Journal of Pharmacognosy and Phytochemical Research* 2, no. 3 (2010): 30–34.

Sawant, R., Binorkar, S.V., Bhoyar, M. and Gangasagre, N.S. "Phytoconstituents bioefficacy and phyto-pharmacological activities of *Terminalia chebula* – A review." *International Journal of Ayurveda & Alternative Medicine* 1, no. 1 (2013): 1–11.

Saxena, M., Faridi, U., Mishra, R., Gupta, M.M., Darokar, M.P., Srivastava, S.K., Singh, D., Luqman, S. and Khanuja, S.P.S. "Cytotoxic agents from *Terminalia arjuna*." *Planta Medica* 73, no. 14 (2007): 1486–1490.

Seguna, L., Singh, S., Sivakumar, P., Sampath, P. and Chandrakasan, G. "Influence of *Terminalia chebula* on dermal wound healing in rats." *Phytotherapy Research* 16, no. 3 (2002): 227–231.

Sekhar, Y.C., Kumar, G.P. and Anilakumar, K.R. "*Terminalia arjuna* bark extract attenuates picrotoxin-induced behavioral changes by activation of serotonergic, dopaminergic, GABAergic and antioxidant systems." *Chinese Journal of Natural Medicines* 15, no. 8 (2017): 584–596.

Senthilkumar, G.P. and Subramanian, S.P. "Biochemical studies on the effect of *Terminalia chebula* on the levels of glycoproteins in streptozotocin-induced experimental diabetes in rats." *Journal of Applied Biomedicine* 6 (2008): 105–115.

Seo, J.B., Jeong, J.Y., Park, J.Y., Jun, E.M., Lee, S.I., Choe, S.S., Park, D.Y., Choi, E.W., Seen, D.S., Lim, J.S. and Lee, T.G. "Anti-arthritic and analgesic effect of NDI10218, a standardized extract of *Terminalia chebula*, on arthritis and pain model." *Biomolecules and Therapeutics (Seoul)* 20, no. 1 (2012): 104–112.

Shaila, H.P., Udupa, A.L. and Udupa, S.L. "Preventive actions of *Terminalia belerica* in experimentally induced atherosclerosis." *International Journal of Cardiology* 49, no. 2 (1995): 101–106.

Shaila, H.P., Udupa, A.L. and Udupa, S.L. "Hypolipidemic activity of three indigenous drugs in experimentally induced atherosclerosis." *International Journal of Cardiology* 67, no. 2 (1998): 119–124.

Sharma, A. and Mukundan, U. "Anti-obesity and antiphyperlipidemic activity of *Terminalia catappa* Linn. in high-fat-diet induced obese rats." *Asian Journal of Pharmaceutical Research and Development* 1, no. 6 (2013): 114–120.

Sharma, M., Lobo, R., Setty, M.M., Saleemulla, K., Chandrashekhar, K.S. and Sreedhara, C.S. "Free radical scavenging potential of *Terminalia tomentosa* (roxb.) bark-an in vitro study." *World Journal of Pharmaceutical Research* 2, no. 6 (2013): 2373–2381.

Sharma, P., Prakash, T., Kotresha, D., Ansari, M.A., Sahrm, U.R., Kumar, B., Debnath, J. and Goli, D. "Antiulcerogenic activity of *Terminalia chebula* fruit in experimentally induced ulcer in rats." *Pharmaceutical Biology* 49, no. 3 (2011): 262–268.

Sheng, Z., Zhao, J., Muhammad, I. and Zhang, Y. "Optimization of total phenolic content from *Terminalia chebula* Retz. fruits using response surface methodology and evaluation of their antioxidant activities." *PLoS One* 13, no. 8 (2018): e0202368.

Shinde, S.L., Junne, S.B., Wadje, S.S. and Baig, M.M.V. "The diversity of antibacterial compounds of Terminalia species (Combretaceae)." *Pakistan Journal of Biological Sciences* 12, no. 22 (2009): 1483–1486.

Singh, A., Bajpai, V., Kumar, S., Arya, K.R., Sharma, K.R. and Kumar, B. "Quantitative determination of isoquinoline alkaloids and chlorogenic acid in Berberis species using ultrahigh performance liquid chromatography with hybrid triple quadrupole linear ion trap mass spectrometry." *Journal of Separation Science* 38, no. 12 (2015): 2007–2013.

Singh, A., Bajpai, V., Kumar, S., Kumar, B., Srivastava, M. and Rameshkumar, K.B. "Comparative profiling of phenolic compounds from different plant parts of six Terminalia species by liquid chromatography–tandem mass spectrometry with chemometric analysis." *Industrial Crops and Products* 87 (2016b): 236–246.

Singh, A., Bajpai, V., Kumar, S., Sharma, K.R. and Kumar, B. "Profiling of gallic and ellagic acid derivatives in different plant parts of *Terminalia arjuna* by HPLC-ESI-QTOF-MS/MS." *Natural Product Communications* 11, no. 2 (2016a): 239–244.

Singh, M.P., Gupta, A. and Sisodia, S.S. "Ethno and modern pharmacological profile of Baheda (Terminalia bellerica): A review." *The Pharmaceutical and Chemical Journal* 5, no. 1 (2018a): 153–162.

Singh, G. and Kumar, P. "Extraction, gas chromatography-mass spectrometry analysis and screening of fruits of *Terminalia chebula* Retz. for its antimicrobial potential." *Pharmacognosy Research* 5, no. 3 (2013): 162–168.

Singh, A., Kumar, S. and Kumar, B. "LC-MS identification of proanthocyanidins in bark and fruit of six Terminalia species." *Natural Product Communication* 13, no. 5 (2018b): 555–560.

Singh, P. and Malhotra, H. "Terminalia Chebula: A review pharmacognistic and phytochemical studies." *International Journal of Recent Scientific Research* 8, no. 11 (2017): 21496–21507.

Singh, D.V., Verma, R.K., Gupta, M.M. and Kumar, S. "Quantitative determinationof oleane derivatives in *Terminalia arjuna* by high performance thin layerchromatography." *Phytochemical Analysis* 13, no. 4 (2002): 207–210.

Smith, N., Mori, S.A., Henderson, A., Stevenson, D.W. and Heald, S.V. *Flowering Plants of the Neotropics*. Princeton: Princeton University Press, 2004.

Sobeh, M., Mahmoud, M.F., Hasan, R.A., Abdelfattah, M.A., Osman, S., Rashid, H.O., El-Shazly, A.M. and Wink, M. "Chemical composition, antioxidant and hepatoprotective activities of methanol extracts from leaves of *Terminalia bellirica* and *Terminalia sericea* (Combretaceae)." *PeerJ* 7 (2019): e6322. doi:10.7717/peerj.6322.

Sornwatana, T., Bangphoomi, K., Roytrakul, S., Wetprasit, N., Choowongkomon, K. and Ratanapo, S. "Chebulin: Terminalia chebula Retz. fruit-derived peptide with angiotensin-I-converting enzyme inhibitory activity." *Biotechnology and Applied Biochemistry* 62, no. 6 (2015): 746–753.

Srinivasan, R., Dhanalekshmi, U.M., Gowri, T. and Duarah, S. "*Terminalia paniculata* bark extract for antidiabetic activity." *International Journal of Pharmaceutical Science and Research* 7, no. 3 (2016): 1331–1337.

Srivastava, S.K., Srivastava, S.D. and Chouksey, B.K. New antifungal constituents from *Terminalia alata. Fitoterapia* 72, no. 2 (2001): 106–112.

Suchalatha, S. and Devi, C.S. "Protective effect of *Terminalia chebula* against experimental myocardial injury induced by isoproterenol." *Indian Journal of Experimental Biology* 42, no. 2 (2004): 174–178.

Sultana, S., Ali, M., Mir, S.R. and Iqbal, D. "Isolatuin and characterization of glycosides from *Convolvulus prostratus, Ficus virens, Phoenix dactifera, Spondias mangifera* and *Terminalia belerica*." *European Journal of Pharmaceutical and Medical Research* 5, no. 1 (2018): 310–318.

Taamalli, A., Iswaldi, I., Arráez-Román, D., Segura-Carretero, A., Fernández-Gutiérrez, A. and Zarrouk, M. "UPLC–QTOF/MS for a rapid characterisation of phenolic compounds from leaves of *Myrtus communis* L." *Phytochemical Analysis* 25, no. 1 (2014): 89–96.

Tala, V.R.S., Candida da Silva, V., Rodrigues, C.M., Nkengfack, A.E., Campaner dos Santos, L. and Vilegas, W. "Characterization of proanthocyanidins from *Parkia biglobosa* (Jacq.) G. Don. (Fabaceae) by flow injection analysis – Electrospray ionization ion trap tandem mass spectrometry and liquid chromatography/ electrospray ionization mass spectrometry." *Molecules* 18, no. 3 (2013): 2803–2820.

Talwar, S., Nandakumar, K., Nayak, P.G., Bansal, P., Mudgal, J., Mor, V., Rao, C.M. and Lobo, R. "Anti-inflammatory activity of *Terminalia paniculata* bark extract against acute and chronic inflammation in rats." *Journal of Ethnopharmacology* 134, no. 2 (2011): 323–328.

Tanaka, M., Kishimoto, Y., Saita, E., Suzuki-Sugihara, N., Kamiya, T., Taguchi, C., Iida, K. and Kondo, K. "*Terminalia bellirica* extract inhibits low-density lipoprotein oxidation and macrophage inflammatory response *in vitro*." *Antioxidants* 5, no. 2 (2016): 20. doi:10.3390/antiox5020020.

Tarasiuk, A., Mosińska, P. and Fichna, J. "Triphala: Current applications and new perspectives on the treatment of functional gastrointestinal disorders." *Chinese Medicine* 13 (2018): 1–11. https://doi.org/10.1186/s13020-018-0197-6.

Thomas, J., Joy, P.P., Mathew, G., Skaria, S., Duethi, B.P. and Joseph, T.S. *Agronomic Practices for Aromatic and Medicinal Plant*. Calicut: Directorate of Arecanut and Spices Development India. Calicut, Kerala, 2000: 124–128.

Thomson, L.A.J. and Evans, B. "*Terminalia catappa* (tropical almond), ver. 2.2." In: Elevitch, C.R. (ed.). *Species Profiles for Pacific Island Agroforestry*. Hōlualoa: Permanent Agriculture Resources (PAR), Hawai'i, USA 2006: 1–20. http://www. traditionaltree.org.

Upadhyay, A., Agrahari, P. and Singh, D.K. "A review on the pharmacological aspects of *Terminalia chebula*." *International Journal of Pharmacology* 10, no. 6 (2014): 289–298.

Uthirapathy, S. and Ahamad, J. "Phytochemical analysis of different fractions of *Terminalia arjuna* bark by GC-MS." *International Research Journal of Pharmacy* 10, no. 1 (2019): 42–48.

Vahab Abdul, A. and Harindran, J. "Hepatoprotective activity of bark extracts of *Terminalia catappa* Linn in albino rats." *World Journal of Pharmacy and Pharmaceutical Science* 5, no. 6 (2016): 1001–1016.

Valsaraj, R., Pushpangadan, P., Smitt, U.W., Adsersen, A., Christensen, S.B., Sittie, A., Nyman, U., Nielsen, C. and Olsen, C.E. "New Anti-HIV-1, antimalarial, and antifungal compounds from *Terminalia bellerica*." *Journal of Natural Products* 60, no. 7 (1997): 739–742.

Venkatalakshmi, P., Vadivel, V. and Brindha, P. "Phytopharmacological significance of *Terminalia catappa* L.: An updated review." *An International Journal of Research in Ayurveda and Pharmacy* 7, no. S2 (2016a): 130–137.

Venkatalakshmi, P., Vadivel, V. and Brindha, P. "Identification of flavonoids in differentp of *Terminalia catappa* L. using LC-ESI-MS/MS and investigation of their anticancer effect in EAC cell line model." *Journal of Pharmaceutical Sciences and Research* 8, no. 4 (2016b): 176–183.

Vonshak, B.O., Sathiyomoorthy, P., Shalev, R., Vardy, D. and Golan, G.A. "Screening of South- Indian medicinal plants for anti-fungal activity." *Physiotherapy Research* 17, no. 9 (2003): 1123–1125.

Walia, H. and Arora, S. "Terminalia chebula-A pharmacognistic account." *Journal of Medicinal Plants Research* 7, no. 20 (2013): 1351–1361.

Wan, C.X., Luo, J.G., Gu, Y.C., Xu, D.R. and Kong, L.Y. "Characterization of homo-flavonoids from three Ophioglossum Species using liquid chromatography with diode array detection and electrospray ionisation tandem mass spectrometry." *Phytochemical Analysis* 24, no. 6 (2013): 541–549.

Warrier, P.K., Nambiar, V.P.K. and Ramankutty, C. *Indian Medicinal Plants. A Compendium of 500 Species* (Vols. 1–5). Madras: Orient Longman Pvt. Ltd., India, 1993.

Wen, K.C., Shih, I., Hu, J.C., Liao, S.T., Su, T.W. and Chiang, H.M. "Inhibitory effects of *Terminalia catappa* on UVB-induced photodamage in fibroblast cell line." *Evidence Based Complementary and Alternative Medicine* 2011 (2010): 904532.

Wright, M.H., Courtney, R., Greene, A.C. and Cock, I.E. "Growth inhibitory activity of Indian Terminalia spp. against the zoonotic bacterium *Bacillus anthracis*." *Pharmacognosy Communications* 6, no. 1 (2016b): 2–9.

Wright, M.H., Greene, A.C. and Cock, I.E. "Investigating the pharmacognostic potential of Indian Terminalia Spp. in the treatment and prevention of Yersiniosis." *Pharmacognosy Communications* 7, no. 3 (2017): 108–113.

Wright, M.H., Jean Arnold, M.S., Lee, C.J., Courtney, R., Greene, A.C. and Cock, I.E. "Qualitative phytochemical analysis and antibacterial activity evaluation of Indian Terminalia spp. against the pharyngitis causing pathogen Streptococcus pyogenes." *Pharmacognosy Communications* 6, no. 2 (2016a): 85–92.

Yadava, R.N. and Rathore, K. "A new cardenolide from the seeds of *Terminalia arjuna* (W&A)." *Journal of Asian Natural Products Research* 2, no. 2 (2000): 97–101.

Yadava, R.N. and Rathore, K. "A new cardenolide from the seeds of *Terminalia bellerica*." *Fitoterapia* 72, no. 3 (2001): 310–312.

Yang, S.F., Chen, M.K., Hsieh, Y.S., Yang, J.S., Zavras, A.I., Hsieh, Y.H., Su, S.C., Kao, T.Y., Chen, P.N. and Chu, S.C. "Anti metastatic effects of *Terminalia catappa* L. on oral cancer via a down-regulation of metastasis-associated proteases." *Food and Chemical Toxicology* 48, no. 4 (2010): 1052–1058.

Yang, B., Kortesniemi, M., Liu, P., Karonen, M. and Salminen, J.P. "Analysis of hydrolyzable tannins and other phenolic compounds in emblic leaf flower (*Phyllanthus emblica* L.) fruits by high performance liquid chromatography-electrospray ionization mass spectrometry." *Journal of Agricultural and Food Chemistry* 60, no. 35 (2012): 8672–8683.

Yuzuak, S., Ballington, J. and Xie, D.Y. "HPLC-qTOF-MS/MS-based profiling of flavan-3-ols and dimeric proanthocyanidins in berries of two Muscadine grape hybrids FLH 13-11 and FLH 17–66." *Metabolites* 8, no. 4 (2018): 57. doi:10.3390/metabo8040057.

Zhang, X.R., Kaunda, J.S., Zhu, H.T., Wang, D., Yang, C.R. and Zhang, Y.J. "The genus Terminalia (Combretaceae): An ethnopharmacological, phytochemical and pharmacological review." *Natural Products and Bioprospecting* 9, no. 6 (2019): 357–392.

Zhang, Y., Liu, X., Gao, S., Qian, K., Liu, Q. and Yin, X. "Research on the Neuroprotective Compounds in *Terminalia chebula* Retz Extracts in-vivo by UPLC–QTOF-MS." *Acta Chromatographica* 30, no. 3 (2018): 169–174.

Zhao, Y., Liu, F., Liu, Y., Zhou, D., Dai, Q. and Liu, S. "Anti-arthritic effect of chebulanin on collagen-induced arthritis in mice." *PLoS ONE* 10, no. 9 (2015): e0139052. https://doi.org/10.1371/journal.pone.0139052.

Zhou, H., Xing, J., Liu, S., Song, F., Cai, Z., Pi, Z., Liu, Z. and Liu, S. "Screening and determination for potential α-glucosidase inhibitors from leaves of *Acanthopanax senticosus* Harms by using UF-LC/MS and ESI-MS[n]." *Phytochemical Analysis* 23, no. 4 (2012): 315–323.

Zywicki, B., Reemtsma, T. and Jekel, M. "Analysis of commercial vegetable tanning agents by reversed-phase liquid chromatography-electrospray ionization-tandem mass spectrometry and its application to waste water." *Journal of Chromatography A* 970, no. 1–2 (2002): 191–200.

Index

Printed in the United States
by Baker & Taylor Publisher Services

.